Consumer Packaging Strategy

The consumer packaged goods (CPG) industry is dominated by major Western brands. The dominance of such major brands extends to burgeoning Asian markets. These conglomerates often rely on packaging as a strategic tool to entice Asian consumers. This book illustrates how packaging as a marketing tool is more than simply changing the label or translating the brand into vernacular language. It examines how different packaging elements (e.g. information, imagery, packaging type) can help to communicate product values to Asian consumers.

Drawing upon rich knowledge of the Asian CPG markets with extensive findings from fieldworks in the key Asian markets, this book explains how Western brands are localising their packaging design in Asian markets. It provides invaluable insight into how major Western CPG brands have relied heavily on their packaging strategies to compete not only against domestic brands but also against other foreign brands.

The book includes in-depth interviews with brand managers of several major Western CPG brands and retailers, and sheds light on emerging trends of CPG packaging in Asia.

Huda Khan is Senior Lecturer in Marketing and Associate Director of Africa Asia Centre of Sustainability Research at the University of Aberdeen, UK.

Richard Lee is Associate Professor of Marketing at the University of South Australia and Senior Research Associate at the Ehrenberg-Bass Institute.

Polymeros Chrysochou is Professor of Marketing at the Department of Management, Aarhus University, Denmark.

Routledge Advances in Management and Business Studies

Halal Logistics and Supply Chain Management
Recent Trends and Issues
Edited by Nor Aida Abdul Rahman, Azizul Hassan and Hajjah Zawiah Abdul Majid

Sustainable International Business Models in a Digitally Transforming World
Edited by Anshuman Khare, Arto Ojala and William W. Baber

Cross-Cultural Challenges of Managing 'One Belt One Road' Projects
The Experience of the China-Pakistan Economic Corridor
Arshia Mukhtar, Ying Zhu, You-il Lee, Mary Bambacas and S. Tamer Cavusgil

Digital Transformation and Innovation in Tourism Events
Edited by Azizul Hassan

Reframing Mergers and Acquisitions around Stakeholder Relationships
Economic, Political and Social Processes
Simon Segal, James Guthrie and John Dumay

Managing Manufacturing Knowledge in Europe in the Era of Industry 4.0
Justyna Patalas-Maliszewska

Family Business and Management
Objectives, Theory, and Practice
Magdalena Biel and Beata Slusarczyk

Consumer Packaging Strategy
Localisation in Asian Markets
Huda Khan, Richard Lee and Polymeros Chrysochou

For more information about this series, please visit: www.routledge.com/Routledge-Advances-in-Management-and-Business-Studies/book-series/SE0305

Consumer Packaging Strategy

Localisation in Asian Markets

Huda Khan, Richard Lee and Polymeros Chrysochou

Routledge
Taylor & Francis Group
LONDON AND NEW YORK

First published 2023
by Routledge
4 Park Square, Milton Park, Abingdon, Oxon OX14 4RN

and by Routledge
605 Third Avenue, New York, NY 10158

Routledge is an imprint of the Taylor & Francis Group, an informa business

British Library Cataloguing-in-Publication Data
A catalogue record for this book is available from the British Library

Library of Congress Cataloging-in-Publication Data
Names: Khan, Huda, author. | Lee, Richard (Professor of marketing) author. |
Chrysochou, Polymeros, author.
Title: Consumer packaging strategy : localisation in Asian markets /
HudaKhan, Richard Lee and Polymeros Chrysochou.
Description: 1 Edition. | New York, NY : Routledge, 2023. |
Series: Routledge advances in management and business studies |
Includes bibliographical references and index.
Identifiers: LCCN 2022009634 (print) | LCCN 2022009635 (ebook) |
ISBN 9780367430368 (hardback) | ISBN 9781032324036 (paperback) |
ISBN 9781003000822 (ebook)
Subjects: LCSH: Packaging–Asia. | Packaging–Asia–Design. |
Branding(Marketing)–Management. | Business planning–Asia. |
Consumer satisfaction–Asia.
Classification: LCC HF5770 .K43 2023 (print) | LCC HF5770 (ebook) |
DDC338.4/76888–dc23/eng/20220331
LC record available at https://lccn.loc.gov/2022009634
LC ebook record available at https://lccn.loc.gov/2022009635

ISBN: 978-0-367-43036-8 (hbk)
ISBN: 978-1-032-32403-6 (pbk)
ISBN: 978-1-003-00082-2 (ebk)

DOI: 10.4324/9781003000822

Typeset in Galliard
by Newgen Publishing UK

Contents

Figures

Tables

Contributors

Susan Freeman

Susan Freeman holds a Ph.D. in international marketing/international business; M.Ed.Sts specialising in assessment and curriculum design; B.Eco (Monash University); and Dip.Ed in Ethics/International Business/Economics/Law (Australian Catholic University, former Institute of Catholic education), Australia. Her research focus is international business and entrepreneurship strategy (venture creation, development and international growth) for smaller firms and MNE; services internationalization; CSR, ethics and global business; in emerging and transitional economies, with an interest in the Asian region, and qualitative research. She was appointed Professor of International Business and Dean: Postgraduate and Professional Programs, University of South Australia, UniSA Business School in 2015. She was appointed Discipline Leader in International Business and Associate Director of the Australian Centre for Asian Business in 2016. In 2016, she founded, and is Director of Strategy, International Business and Entrepreneurship Research in the Business School.

Anne Sharp

Associate Professor Anne Sharp has been a senior marketing scientist with the Ehrenberg-Bass Institute, at the University of South Australia. Anne has a particular interest in sustainable marketing and interventions encouraging behaviour change for improved environmental outcomes. Her recent work includes bans on single use plastics and plastic bags, waste and recycling behaviour, and the sharing and circular economies. With over 50 academic papers and book chapters, Anne is a full member of the Australian Market and Social Research Society.

Wayne Wang

Wenjun (Wayne) Wang is the Head of International Business at Playgro International—a world-leading infant developmental toy brand. Wayne has previously worked in managerial positions in several Fortune 500 corporations across North America and the Asia Pacific regions, including Staples, 3M and

Newell. He is specialised in branding, marketing, business development and new product innovation. Wayne moved to Australia in 2016 and holds a Master by Research degree from the University of South Australia.

Song Yang

Song Yang is a marketing scientist at Ehrenberg-Bass Institute for Marketing Science at University of South Australia. He holds a Bachelor of Economics and a Master of Economics from Renmin University of China in Beijing, and a Ph.D. in Marketing from the University of Western Australia. He has published many articles widely in academic journals and his research interests include consumer behaviour under the cultural influence, consumer value changes in China, research methodology in marketing and cross-cultural management.

Abbreviations

AIDA	attention, interest, desire and action
CPG	consumer packaged goods
DCE	discrete choice experiment
EEG	electroencephalography
fMRI	functional magnetic resonance imaging
GDP	gross domestic product
NFA	need for affect
NFC	need for cognition
P&G	Procter & Gamble

1 The Consumer Packaged Goods Industry in Asia

Growth of Asian Consumer Packaged Goods Markets

The trading and movement of products around the globe is anything but new. Since the ancient Silk Road that connected the East and West, or the excavated Suez Canal that opened up an artificial waterway between the North Atlantic and the Indian Ocean, products have moved from one country to others in a complex web of global commerce. The advent of the internet and e-commerce has even led to the instantaneous transfer of intangible products. While the West—first, the United States and later Europe—has largely driven global trade after World War II, by the 21st century, there is little doubt that the East is firmly established as the centre of economic gravity and a critical driver of global growth. The United Nations Department of Economic and Social Affairs (UN DESA, 2019) reported that the global population was about 7.7 billion in 2019. Asia accounted for about 60% or 4.6 billion of this figure, and its population is expected to reach 5.3 billion by 2050. While Asia accounted for less than one-third of the global gross domestic product (GDP; in terms of purchasing power parity (PPP)) in 2000, it is on track to contribute about half by 2040.

Underpinned by rapid strides in human development, economic progress that lifts millions out of poverty, longer life expectancy and greater technological literacy, Asian consumers are projected to foot 40% of the world's consumption expenses by 2040 (Tonby, Woetzel, Choi, Seong, & Wang, 2019). Estimates also suggest that by the end of 2030, about 60% of households with annual income exceeding US$22,000 will be based in Asia (Tonby et al., 2021). Although pockets of poverty and uneven development challenges remain in some countries or intra-country regions, the overall rise in living standards and income level, coupled with increasing urbanisation, makes Asian consumers a key target for many marketers.

A key driving factor behind Asia's GDP growth and soaring consumption is the dramatic rise of its middle class (Tonby et al., 2019; Woetzel, Tonby, Thompson, Burtt, & Lee, 2014). Even the smaller Southeast Asian region alone is anticipated to experience a doubling of its middle class from about 80 million households to more than 160 million by 2030 (Woetzel et al., 2014). Further, the 2019 World Economic Forum has forecasted that by 2030, an overwhelming

DOI: 10.4324/9781003000822-1

majority (90%) of the 2.4 billion new middle-class consumers entering the global economy will hail from Asia-Pacific (Yendamuri & Ingilizian, 2019). A recent report by the McKinsey Global Institute (Tonby et al., 2019) acknowledges that the 'world's future' has arrived sooner than many realise. Indeed, Asia's Generation Z has grown up with unprecedented wealth. Coupled with greater exposure to Western culture, these young, affluent and digital-savvy Asian consumers are heavily influenced by their media consumption and their desire for conspicuous consumption.

Take China and Pakistan, two Asian countries that we cover in greater depth in later chapters in this book. China's domestic economy has doubled about every seven years between the 1970s and early 2000s, with economic growth averaging about 10% per annum during this period (Lysonski, 2014). According to public data from the World Bank, China's GDP in 2019 was approximately US$14.3 trillion, or about 16% of the world's GDP. Although China's GDP growth of 6% in 2018–2019 was the lowest since 1991, it still compared favourably with the global average GDP growth of 2.6% for the same period. The International Monetary Fund reported that on PPP basis, China has overtaken the United States as the world's largest economy since 2014 (Media, 2014). Recent releases of key economic data by China reveal that the country has continued to power on economically despite the woes brought about by the COVID-19 pandemic (Statistica, 2021).

Along with this growth, China is experiencing a rapid expansion of its middle-class population. By the end of this decade, it will exceed 1.2 billion, comprising one-quarter of the world's total middle-class population. This growth mirrors a McKinsey report, which estimated that 730 million urban Chinese fell into the middle-class income categories of 'aspirants' or 'affluents' in 2016 (Ho, Felix Poh, Jia Zhou, & Zipster, 2019). By 2030, China's working population alone is projected to account for 12 cents of every US$1 of worldwide urban consumption (Tonby et al., 2019).

Although the absolute numbers for Pakistan pale in comparison with those for China, developing countries such as Pakistan have also contributed towards Asia garnering a greater slice of the world's GDP (Dobbs et al., 2012; Farrell, Gersch, & Stephenson, 2006). Pakistan's economy, which grew 5.8% in 2018 to about US$315 billion, has witnessed dramatic year-on-year growth since 2000 when its GDP was about US$83 billion. On average, the country's GDP has increased about 5% per annum between 2005 and 2014 (Trading-Economics, 2015). As is the case with China, the factors contributing to Pakistan's GDP growth are its increasing urbanisation and the rise of its middle class (Pakistan Economic Survey, 2015). A population census similarly revealed that the Pakistani middle class is expanding rapidly. Annual disposable income per household has increased by 143% during 1995–2011, and this trend of growth is likely to persist well into the future (Euromonitor, 2013).

Given the population growth in Asia, a burgeoning middle-class segment, rapid urbanisation and increasing income level, one particular industry stands to benefit substantially—the *consumer packaged goods* (CPG) industry (Razdan,

Yamakawa, Francois, & Seong, 2021). As affluence increases, people's consumption naturally increases as well. Asia's consumption of CPG has nearly doubled from US$32 billion in 1997 to US$63 billion in 2017 (Tonby et al., 2019). The increase in the number of retail outlets catering to these consumers in urban areas, as well as the proliferation of online shopping, further facilitates CPG consumption. With over 30 million retail stores across Asia (compared with about 1 million in the United States), the region's share of the global CPG industry will exceed US$750 billion, that is, about 35% (Accenture Consulting, 2018).

As the Asian consumer class attains higher income levels, thereby shifting the income pyramid sharply upwards, markets will also see greater discretionary spending and brand premiumisation. As established Western brands reach maturity in their markets of operation, a common business practice they adopt is seeking growth opportunities in other markets (Murray, Gao, Kotabe, & Zhou, 2007). Asia has thus attracted major Western CPG brands, such as Procter & Gamble (P&G), Nestle and Coca-Cola (Gupta & Wright, 2019). The rapidly growing economies in this region, and the size and consumption power of the middle class, make Asia a strategically and financially important target for many Western CPG brands (He & Wang, 2017; Zhou & Hui, 2003).

In particular, the rise in individual purchasing power has led to China being the fastest growing and most important market for many CPG brands. As they prosper, Chinese consumers demand more Western brands (Cui & Liu, 2001; Durvasula & Lysonski, 2010; Guo, 2013). For example, 90% of Chinese families have chosen P&G products, and other brands (e.g. Mars, Colgate, Coca-Cola and Unilever) have also successfully achieved more than 70% market penetration in 2017 (China Internet Watch, 2017). Tai and Pae (2002) similarly found that foreign brands have garnered strong recognition among Chinese consumers in comparison to local brands, with nine out of the top ten brand awareness leaders in China being foreign brands, such as Coca-Cola.

Similarly, a recent economic survey shows that Pakistani consumers' literacy and awareness of, as well as demand for, Western brands are persistently strong (Pakistan Economic Survey, 2015). They are consumption oriented and spend significantly on consuming Western brands (Khan & Bamber, 2007). This fact explains Pakistan's substantial expenditure on imported goods, particularly by its middle and upper classes, which is estimated to be approximately $10 billion per annum (Khan & Bamber, 2007). These consumers contribute to about 30% of the country's total domestic expenditure, with the highest expenditure on foreign brands (Trading-Economics, 2015). A report on CPG consumption in Pakistan clearly indicates a shift in consumer preference from traditional loose or unpackaged foods to packaged products, mainly for the reasons of quality, safety and hygiene (Euromonitor, 2015). The shift to packaged products is also because Pakistani consumers are becoming more brand conscious. For example, instead of consuming cow milk collected and bottled by farmers, they have shifted to purchasing packaged milk, particularly those of Western origin.

Packaging Localisation as a Marketing Strategy

Asia is, no doubt, crucial to Western CPG brands' drive to increase their share of the global CPG market. Breaking into the region, however, poses significant challenges to brand managers due to the region's geographic and ethnic diversity. For example, in a recent McKinsey report regarding the CPG industry in Asia, Razdan et al. (2021) highlighted that changes in demographics, such as the shrinking household size, an ageing population in some markets, an increasingly voracious digital generation and a greater awareness of issues such as sustainability and ethics, require brand managers to rethink their marketing strategies in order to compete successfully, not only against other major foreign brands but also against emerging or dominant home-grown brands.

Another challenge is the emerging, or even dominant, local brands in Asian markets including China, Indonesia, South Korea and Vietnam (Kantar, 2019). Despite the rising income levels and the increased foreign CPG consumption in these countries, local brands are taking market share from established Western brands. Local players' ability to adapt quickly to local consumer trends has enabled them to launch innovative products or marketing initiatives into the local market more effectively than established Western brands, which have relied mostly on identical products in their home markets. For example, Indomie is the most chosen instant noodle brand in Indonesia, far outselling major brands such as Maggi. Its success rests on appealing to the taste familiar to local residents. Similarly, Liby, a market-leading Chinese washing detergent that outsells global brands such as Unilever's OMO (Statistica, 2019), uses local celebrities with a clean, modern image to endorse its products.

In light of these challenges, managers of major Western CPG brands have relied on a suite of localisation strategies to attract Asian consumers. The idea is simply that instead of adopting a standard approach by using a uniform marketing mix (price, product, distribution and promotion) to enact a similar strategy across countries (Chiou & Droge, 2015; Kravets & Sandikci, 2014; Sheth, 2011), brands would tailor marketing strategies according to the distinctive differences within each market in order to appeal to consumers in the market (Bai, McColl, & Moore, 2021; Khan & Lee, 2020a; Rigby & Vishwanath, 2006). These differences include culture (Busnaina, Almgatif, & Sdiri, 2021; De Mooij, 2010; Hofstede, 2001), language (Ho, Chiu, Jiang, Shen, & Xu, 2019; Khan & Lee, 2020a, 2020b), economic development stage (Hoskisson, Eden, Lau, & Wright, 2000) and industry structure (Paliwoda, Slater, Vrontis, Thrassou, & Lamprianou, 2009; Vrontis, 2005). Consistent with social identity theory (Deshpandé & Stayman, 1994), the rationale is that domestic consumers would identify with products that they perceive as local and hence would be more inclined to favour and buy them (Poon, Evangelista, & Albaum, 2010).

Localisation marketing strategies include ensuring advertisement content is more relevant to local audiences (Hornikx & Meurs, 2021; Micu & Coulter, 2010), changing product content to cater to local tastes or requirements (Kantar,

2019) and using local celebrities for endorsements (Martin & Lee, 2004). Numerous studies have shown these localisation strategies are effective in influencing consumers. For example, ethnic models in advertisements could boost ethnic viewers' self-referencing, thereby leading to improved attitudes towards, and intentions to buy, the advertised products (Martin & Lee, 2004). Similarly, Deshpandé and Stayman (1994) show that exposing consumers to advertising cues related to their culture improved their perceptions of the advertisements. Creating marketing messages in a native language that is congruent with individuals' social identity would also create favourable perceptions of the advertisement and the advertised products. This view is consistent with the assertion that people tend to favour things related to their own social group, rather than those of groups from a dissimilar social background (Turner, Brown, & Tajfel, 1979). Adapting marketing practices is particularly crucial because of the heterogeneity of consumer behaviour across a market as diverse as Asia (Ozturk & Cavusgil, 2019). The cultural differences in this region are so vast that it makes standardisation problematic and unfeasible, leading to the loss of competitive advantage (Melewar & Vemmervik, 2004; Yap & Yazdanifard, 2014). By adapting marketing strategies to each target country, localisation can tap the uniqueness of each market in order to differentiate from, and gain advantages against, competing brands (Bai et al., 2021).

Another effective marketing strategy that foreign CPG brands use in Asian markets is to customise their product packaging to enable local consumers to understand, recognise and relate to the products better (Khan & Lee, 2020a; Khan, Lockshin, Lee, & Corsi, 2017). Packaging localisation is particularly important because, as a strategic marketing tool, packaging entails several different elements (e.g. language, colour, labelling and price) that play important roles in communicating product information, creating value (Lisa, 2007; Underwood, 2003), shaping quality perceptions (Aprile, Caputo, & Nayga, 2012), affecting the experience of contents (Khan & Lee, 2020b; Velasco & Spence, 2018) and, ultimately, biasing consumer choice selection (Khan, Lockshin, et al., 2017; Silayoi & Speece, 2004). These packaging elements offer opportunities for brands to customise their packaging in different ways in order to target local consumers more effectively. Indeed, some researchers have even argued that the influence of packaging on consumer behaviour occurs below conscious awareness (Spence, Velasco, & Petit, 2019; Sugermeyer, 2021).

> This book focuses on a common and effective marketing strategy used by Western CPG brands in Asian markets—that of localising the packaging of the products that these brands export to, or manufacture in, Asia for Asian consumers.

Despite the importance of packaging as a marketing communication tool, Western CPG companies still allocate the majority of their media budget, estimated at about 70%, to TV advertising alone (Accenture Consulting, 2018). Another striking point is that if CPG companies were to increase their TV advertising budget tenfold, they would reach only an additional 1% of potential consumers (Accenture Consulting, 2018). *This situation begets the question whether a larger slice of the budget should go to developing effective localised packaging.* Let us consider China again. Typical supermarkets in China are similar to those in Western countries in that they sell identical products from similar major international brands, such as Mars and P&G. However, a difference regarding supermarkets in China is that the brand names and the descriptions of products are typically localised to the Chinese language. Anecdotal information from informal discussions with shoppers and store personnel in some Chinese supermarkets revealed that they had not seen, or could not recall seeing, major international brands with names and descriptions in the English language.

Similarly, a recent economic survey shows that Pakistani consumers' literacy and awareness of, as well as demand for, Western brands are persistently strong (Kazmi, 2019; Rind, 2015). During our visits to supermarkets in major Pakistani metropolises, we found that the shelves are well stocked with major Western brands, including the likes of Mars, P&G and Unilever. However, the visits also revealed that most Western brands are sold either in the standard packaging that one would typically see in Western markets such as the United States or Australia, or in a localised packaging that is customised for the non-Western market. The images in Figure 1.1 show examples of products in the standard packaging, and the same products in a localised packaging for the Chinese market, in this instance.

This practice of having two different packaging types naturally raises a question: Which one, and under what circumstances, is more effective in attracting local consumers? Indeed, international marketers have continued to grapple with the issue of whether foreign brands should localise their marketing strategies, especially their packaging (Strizhakova & Coulter, 2015). *To this end, this book*

Figure 1.1 Standard packaging versus localised packaging.

Note: 'Coca-Cola Chinese design' by renaissancechambara is licensed under CC BY 2.0.

addresses these questions by exploring, through a series of studies, how packaging localisation may enhance the competitiveness of Western CPG brands in Asian markets.

Structure of This Book

The remainder of this book is organised as follows. After this introductory chapter, Part I of this book begins with Chapter 2, which discusses the role of packaging, beyond its role as a physical container to it being an effective marketing communication tool. The chapter also introduces the concept of packaging elements or cues[1] and draws support from various theories regarding the reasons that packaging elements may shape consumers' perceptions and purchase behaviour of a product via its packaging. Chapter 3 builds on Chapter 2 and presents a holistic versus atomistic perspective to discuss in greater depth the cues of packaging designs, and the consumer responses likely to be elicited by these design elements. Next, Chapter 4 discusses the common methods used by marketers to measure packaging effectiveness when selecting or modifying packaging design, with an emphasis on packaging effectiveness from a cross-cultural perspective. Drawing on the concepts of standardisation and localisation, Chapter 5 sets the theoretical foundation to explain the reasons that localising the packaging of foreign products sold in Asian markets is effective. In doing so, the chapter also describes a case study regarding how major Western CPG brands across multiple categories have localised their products for the China market. In summary, Part I discusses fundamental concepts about packaging, its role in marketing communication, its influence on consumer behaviour and the theory and practice of packaging localisation as a marketing strategy.

The conclusion one can draw from the discussion in Part I relates to the importance of packaging elements in changing the appearance of packaging, and consequently shaping product perceptions by providing or accentuating some type of information or meaning to consumers. However, Burbank and Martins (2010) argue that the interpretation of packaging cues varies across contexts and situations in which consumers comprehend the object. Hence, the associations of cues to meanings are often nuanced or determined by the context within which the product is being marketed (Creusen & Schoormans, 2005). In Part II, this book presents a series of studies to shed light on how the effects of packaging localisation may vary across different contexts and consumption situations. Specifically, Chapter 6 kicks off with a study conducted in Pakistan to show that although packaging elements may influence consumer behaviour, the influence depends on the product type (hedonic versus utilitarian) as well as on the extent of consumer familiarity with the brand. Chapter 7 then replicates the study in Chapter 6, but it adds another key variable known to influence consumer behaviour—product price—and also runs a discrete choice experiment to better mimic a shopping trip. Continuing along the same line of investigation, Chapter 8 reports another study conducted in situ within a supermarket in China. The purpose was to understand how the consumption situation (conspicuous versus inconspicuous consumption) may drive the choice of packaging type, and whether consumers are willing to pay a premium for their choice of packaging under either consumption situation.

Earlier, Chapter 3 discussed how packaging elements may fall into one of three dimensions (informational, graphic or structural). The study reported in Chapter 9 examines how language, a critical and often used element in packaging localisation, may fall into overlapping categories, and the circumstances under which language operates as an informational cue or a graphic cue. Chapter 10 shows that consumers' personal dispositions may also influence their choice of packaging for an identical product. Specifically, it shows that different packaging elements may appeal differently to consumers with need-for-cognition or need-for-affect tendencies.

Thus, the chapters in Part II present studies after exposing participants to actual products found in supermarkets where the studies are located. In these chapters, we primarily demonstrate how the localised versus standard version of a product packaging may elicit different consumer responses. Chapter 11 then concludes Part II of this book by reporting the findings of a series of in-depth interviews conducted with brand managers of major Western CPG brands. The interviews attempt to elucidate the industry's perceived need for packaging localisation as Western brands enter Asian markets, and the rationale behind the current packaging localisation strategies adopted by these brands.

Note

1 We use the term 'packaging element' and 'packaging cue' interchangeably to refer to the individual components or features of a packaging, such as its colour, shape, graphics and brand logos, whereby these elements or cues may individually or collectively influence consumers' perceptions of the product and consequently their purchase behaviour.

Part I

Packaging Design and Localisation Strategy

2 Marketing Role of Packaging

Functions of Packaging

Packaging is typically regarded as a container that holds and protects the product and facilitates handling and commercialisation. Referring to this fundamental role, Rundh (2005) points out that packaging is a container that protects and preserves product content in the distribution process. In his account of the early history of product packaging, Thomas Hine (1995) makes it clear that the original intention of packaging was about holding and preserving products, and hence, the early concerns primarily related to portion control and preservation efficacy. Since the path from manufacturers, via the supply chain, to final consumption may involve transportation around the globe and take several months, packaging plays a key role in fulfilling various functions during this process. In his book on food packaging, Robertson (2005) attributes four functions to packaging: Containment, protection and preservation, convenience, and communication. As much as each of these four functions serves different purposes, each comprises different technological, engineering and commercial objectives (Berk, 2009).

Containment: This is the primary objective. Holding a product appropriately is essential for its efficient transportation, distribution, storage and use. The shape and size of the packaging guide the space requirement not only for storage and transportation, but also for display at the point of sale. The modern emphasis on sustainability further means that the choice of container materials may account for how it will support responsible disposal or recycling, rather than becoming traditional waste.

Protection and preservation: These comprise arguably the most important function of packaging. By shielding the product, particularly food, from contact with the external environment, packaging protects it from potentially harmful physical, chemical or biological contamination. For food, this is a decisive factor in determining the length of product shelf life. Even a superficial imperfection in packaging can lead to negative consumer perceptions about contamination (White, Lin, Dahl, & Ritchie, 2016). Concurrently, the packaging also protects the environment from the product, such as by preventing spillage and odour (think chlorine or bleaching solution). The structural integrity of the packaging

DOI: 10.4324/9781003000822-3

ensures that the product remains undamaged and ready to use as intended. In recent years, modern retailing has influenced packaging owing to changing consumer demands and shifts in the entire supply chain. The modern retail environment has put pressure on retailers to ensure that stores are stocked with fresh supplies of products. This aspect compels retailers to ensure an effective supply chain, especially for fresh products or products with a short shelf life, thus leading to the recognition of packaging as a key preservative tool for the product.

Convenience: Given the heterogeneous requirements of different consumer groups (e.g. household size, consumption rates, portability and delivery), adapting the size of the package to conveniently meet diverse consumers' needs is a critical function of packaging. Ensuring convenience enhances brand competitiveness by allowing brands to serve customers better. Convenience in packaging also concerns using the right type of containers, such as aerosol cans for carpet cleaners, resealable bags for frozen vegetables and cups or bowls from which soup can be heated and consumed directly. These designs provide convenience during product consumption.

Communication of product-related information: Packaging also provides a physical location on which information, including the quantity, country of origin, nutritional data, production or expiry date and barcode are provided. This product-related information serves to facilitate consumer choice making by allowing consumers to evaluate products objectively (Dickson & Sawyer, 1990; Pieters & Warlop, 1999).

> There is yet another function of packaging that concerns this book—its key role as a marketing communication tool to influence consumer perceptions and purchase behaviour.

Packaging as a Marketing Communication Tool

Early research focused on the functionality of packaging design, but scholars soon realised that packaging can also assume a communicative role as a marketing tool (Baccarella, Maier, & Voigt, 2021; Dörnyei, 2020; Khan & Lee, 2020a; Underwood, Klein, & Burke, 2001; Wagner & Charinsarn, 2021). Packaging can act as a powerful brand element to help in communicating product value and persuading consumers to buy (Baruk & Iwanicka, 2015; Oenning, Guimarães Junior, & Medeiros, 2018). Underwood (2003) regarded packaging is a 'brand ambassador' in marketing communications.

Clement, Kristensen, and Grønhaug (2013) reveal that acquiring consumers' attention is the initial, imperative step in shaping their buying decisions. Similarly, a key marketing role of packaging is to capture consumers' notice, often in cluttered retail environments. In this respect, Kapferer (1995) contends that consumers often merely glance at packaging across the shelves in retail stores because they tend to spend limited time and attention on product selection (Pieters & Warlop, 1999). In supermarkets, in particular, each category typically has many brands, and these brands need to compete in drawing consumers' attention in order to deliver the point of sale of the product and the brand. Hence, researchers argue that packaging design should be attractive, and then, it may work as the first point of contact, which can determine consumer preference for the product before they make a choice (Clement et al., 2013; Garber, 1995; Pieters & Warlop, 1999).

Pilditch (1957) likened packaging to a 'silent salesman' that captures consumers' attention and conveys the product's characteristics, such as its quality and benefits, to them. For example, a survey of snack foods found that packaging ranked second behind taste in brand choice criteria (Silayoi & Speece, 2007). The study also reported that 86% of the respondents were willing to pay a higher price for a visually appealing package. Similarly, another study revealed that consumers evaluate message information provided on the package to justify the price charged (Breetz, 2014; Khan, Lockshin, et al., 2017). Indeed, consumers' overall propensity to stay with a brand or switch brands may depend on their perceptions of a product through its packaging (Håkansson, Kvarnström, & Nilsson, 2014). Given its crucial role in influencing consumers, Nickels and Jolson (1976) argue that it should be considered the fifth P of the marketing mix. The growing realisation about the value of packaging design has resulted in a great deal of research on ways to make packaging more effective for brands as well as for consumers.

The efficacious role of packaging as a marketing communication tool can also be gleaned from studies on the retail context, which show that most consumer purchase decisions occur in the store spontaneously at store shelves, rather than being premeditated (Inman, Winer, & Ferraro, 2009; Kacen, Hess, & Walker, 2012). In the consumer packaged goods industry, consumers' decisions are often determined by the information communicated at the point of purchase (Sharp, 2010). The term point of purchase refers to the elements designed to capture the attention of consumers as they are considering their final choice decisions in a category (Silayoi & Speece, 2007; Wertenbroch & Skiera, 2002). Blatt (2002) reports that about 70% of in-store decisions in retail stores are made at the point of purchase. Likewise, Weinberg and Gottwald (1982) find that more than 60% of supermarket buyers and 30% of department store shoppers made their purchase decision based on the point of purchase they experienced in the store. In other words, consumers' intentions to purchase a product depend on the degree to which they expect to derive satisfaction from its use, but when they do not even think much about the product before buying, brand choice decisions are determined by aspects communicated at the point of purchase (Clement et al., 2013; Kacen et al., 2012; Silayoi & Speece, 2007; Steckel et al., 2005). Consumers' decision to choose a particular brand changes when a clutter of information is available at the point of purchase (Steckel et al., 2005). Resonating with this finding, Sharp (2010) contends that unless brands are noticed at the point of purchase, they are unlikely to be purchased. Consequently, packaging has become a crucial factor in marketers' arsenal, for it attempts to shape consumer decision-making at the point in time when they are actually making a choice.

Similarly, the importance of packaging to consumer behaviour is reflected in the amount of time that consumers typically spend noticing products on a typical supermarket shelf. A study found that almost 50% of shoppers spend less than five seconds at the point of purchase (Dickson & Sawyer, 1990). Therefore, as a prominent distinctive brand asset, packaging must be unique and famous enough to be identified quickly by shoppers to ease their choice decisions (Piñero, Lockshin, Kennedy, & Corsi, 2010). Sharp (2010) argues that packaging provides consumers' the ability to recognise the brand, since it communicates the brand identity cues (Nguyen & Gunasti, 2018). Consequently, packaging becomes critical in determining consumers' evaluation and choice based on attention-drawing attributes (Creusen & Schoormans, 2005).

Emerging market trends, such as social media, in-store marketing and advertising, have shifted consumer habits, brand awareness and consumption patterns. These factors have boosted the significance of packaging in influencing consumer behaviour. The most profound reason lies in its attention-drawing ability that results in raising the probability of purchase (Ares, Besio, Giménez, & Deliza, 2010; ., Bagger, Lahm, Grunert, & Scholderer, 2019). Research has established that acquiring consumers' attention is the initial, imperative step in influencing their purchase decisions (Chandon, Hutchinson, Bradlow, & Young, 2009; Clement et al., 2013; Creusen & Schoormans, 2005). Hence, the appearance of packaging is a critical source for grabbing attention; it is the consumers' first point of contact with a product before the actual purchase that often influences product choice (Orth & Crouch, 2014; Pietersen, 2010). Given the clutter of brands that consumers would typically encounter in most product categories, packaging is a tool to help brands stand out and distinguish themselves from competitors.

Packaging Cues

Consumers' expectations and evaluation of a product are often based on its packaging appearance. Consumers may also discern quality from the packaging appearance, thus attesting to the importance of packaging in forming positive or negative expectations about the product (Ragaert, Verbeke, Devlieghere, & Debevere, 2004). If packaging is perceived as inferior quality, the perception of lower quality is also transferred to the product (Silayoi & Speece, 2004). However, what constitutes a good packaging design from a marketing communication perspective? Answering this question is critical because packaging accounts for a substantial amount of investments. Estimates show that approximately 40% of marketing budgets are allocated to packaging designs (Rettie & Brewer, 2000).

When one views a product packaging, one sees multiple sources of information or cues that signal certain characteristics to bias perceptions. As we shall elaborate in the next chapter, cues may be verbal or visual. Language is often used as a verbal cue in marketing communication strategies (Khan & Lee, 2020a). Cues may also be visual. Examples of such cues include the shape of the packaging (think Toblerone or Coca-Cola), the colour or colour scheme (e.g. the shade of blue that is Tiffany's identity) or even visual images (the cartoon characters of M&M's chocolate). Young (2002) lays emphasis on designing effective packaging cues to gain a lead over competing brands. As a specialised form of communication, these verbal or visual cues not only influence consumers at the point of purchase but also continue to communicate brand messages when the products are consumed at home.

Language is a particularly important and efficacious cue (Khan, 2019; Khan, Lee, & Lockshin, 2015; Khan, Lockshin, et al., 2017). For example, a multi-country experiment shows that consumers perceived a product to be more foreign when it was marketed in English (the foreign language in Asia) than when it was marketed in the country's native language (Gerritsen et al., 2007). Steenkamp, Batra, and Alden (2003) argue that the 'foreignness' has strong effects on purchase intentions because of the perceptions about high quality associated with these foreign brands. For example, a study on Chinese consumers found that when the brand name is written in the English language, it enhances the attitude towards the brand (Li, Li, & Kambele, 2012).

Favourable or unfavourable product perceptions may also stem from appealing or non-appealing visual cues present on product packaging, respectively. Such visual cues include colour (Magnier & Schoormans, 2017; Spence & Velasco, 2018), shape (Spence, 2016; van Ooijen, Fransen, Verlegh, & Smit, 2017), size (Arboretti & Bordignon, 2016; Eldesouky, Pulido, & Mesias, 2015) or even religious symbols (Bakar, Khan, Hashim, & Lee, 2021). For example, bright colours on product packaging are visually appealing to some consumers but not to others, and consumers infer the product quality differently because of different meanings attached to colours in different societies (Madden, Hewett, & Roth, 2000). Thus, packaging cue acts as strategic method to enhance consumers' associations with the products, resulting in bias in their preference by developing positive or negative perceptions (Bakar et al., 2021; Håkansson et al., 2014; Silayoi & Speece, 2007).

Feldman and Lynch's (1988) accessibility-diagnosticity theory can be used to explain the influence of packaging cues, since at the point of purchase, packaging is physically more accessible than any other form of marketing communication (Underwood & Klein, 2002). The theory implies that making cues on packaging more accessible (e.g. using the native language familiar to local consumers) would improve diagnosticity and subsequently evaluations by consumers. Consequently, packaging research often applies this theory to explain the influence of packaging information on consumer liking and preference.

Support for the efficacy of packaging cues may also emerge from symbolism theory (Blumer, 1969), which posits that consumers draw certain meaning from cues present on an object, such as its packaging. The association of cues to meanings is often innate or determined by a context where the product is being marketed (Bakar et al., 2021; Creusen & Schoormans, 2005). A classic example of this association is a brand that is perceived as more Western when communication programmes utilise Western brand names or advertisement themes; these cues symbolise Westernisation and modernity for non-Western consumers (Alden, Steenkamp, & Batra, 1999; Batra, Ramaswamy, Alden, Steenkamp, & Ramachander, 2000). This shows that the meaning of a cue on the product

Table 2.1 Selected studies on different cues in packaging

Authors	*Packaging cues*	*Cue type*	*Theory*	*Implications*
Labrecque and Milne (2012)	Colour	Visual	Theories of aesthetics and associative learning	Colours play a role in driving perceptions of personality dimensions of a brand. Combination of colour and logo shape influences familiarity and likeability.
Sample, Hagtvedt, and Brasel (2020)	Shape, colour, materiality, location, illuminance	Visual	Cognitive, perceptual pathways	This systematic review study focuses on visual components of cues and found that five types of visual cues are critical for marketing.
Ikonen, Sotgiu, Aydinli, and Verlegh (2020)	Labels	Verbal	N/A	The meta-analysis study finds that front-of-packaging labels about nutrition nudge consumers towards healthier choices.
Newman, Burton, Andrews, Netemeyer, and Kees (2018)	Labels	Verbal	Dual processing framework	Presence of icons on the front of packaging can decrease intention for unhealthy products. Nutrition consciousness can moderate effects related to consumers' use of back-of-packaging labels.
Zhu, Billeter, and Inman (2012)	Different types, e.g. pictures and brand name	Verbal and visual	Signalling theory	Salient cues in packaging (e.g. pictures and brand names) can increase perceived product efficacy and lead to lower usage amount. The impact of cues on product usage is moderated by need for cognition and cue salience.
	Textual and pictorial	Verbal and visual	Relevance theory	Two types of potentially misleading elements on packaging (textual and pictorial) were found to have analogous effects on consumer preferences.

package is first transferred to the product and consumers purchase the product to benefit from the symbolic value associated with the product via the cue (Rompay & Veltkamp, 2014; Van Ooijen, 2016). Thus, this discussion reveals the importance of cues in changing the appearance of packaging for consumers, which consequently provides symbolic meaning to these consumers.

In summary, packaging creates impressions on consumers by using cues (Baruk & Iwanicka, 2015; Vihma, 1995). Manipulation of even a single cue on product packaging can change its appearance and aesthetics for consumers, and consequently connote different meanings for different consumers (Silayoi & Speece, 2007). Table 2.1 describes some studies on different packaging cues over the past decade that were published in leading marketing journals.

From the review of studies presented in Table 2.1, it can be concluded that packaging cues, both visual and verbal, can have different implications for consumers. A classic example of this is a brand being perceived as foreign when communication programmes utilise foreign brand names or advertisement themes; these cues symbolise Westernisation and modernity for non-Western consumers (Alden et al., 1999; Batra et al., 2000). This finding suggests that the meaning of a cue on the product package first transfers to the product, and consumers purchase the product to benefit from the symbolic meaning or value associated with the product via the cue (Bakar et al., 2021; Rompay & Veltkamp, 2014; Van Ooijen, 2016). Thus, a crucial trend in brand management and consumer behaviour research is to identify the elements of packaging that make it appealing to consumers such that they prefer one brand over a competing brand. To shed further light on this crucial component of effective packaging, the next chapter provides deeper insights into the different types of packaging cues.

3 Conceptualising Packaging Design

Holistic versus Atomistic Perspectives of Packaging Design

A main function of packaging is to communicate product-related information. Designers and marketers often make strategic decisions when choosing packaging elements because, as we have highlighted in Chapter 2, packaging elements are known to influence consumer response, convey values that position the brand in consumers' mind (Orth & Malkewitz, 2008) or even portray stories and imaginations about the brand (Kniazeva & Belk, 2007). Such consumer response consequently affects the purchase intentions regarding products. Figure 3.1 depicts this relationship schematically.

The ways in which consumers respond to packaging design depends on the way they process packaging information. In the literature, two main perspectives on how consumers process packaging information have been discussed: the *atomistic* and the *holistic* perspectives. Homburg, Schwemmle, and Kuehnl (2015, p. 42) describe these two perspectives as 'gestalt processing on the gestalt level and atomistic processing on the parts level'.

The *atomistic* perspective is rooted in theories of visual rhetoric and semiotics. Research following this perspective examines the impact of individual packaging cues (e.g. colour, imagery, material and typeface) on consumer response. The literature grounded in this perspective shows that individual packaging design elements, such as colours (Baptista, Valentin, Saldaña, & Behrens, 2021; Becker, van Rompay, Schifferstein, & Galetzka, 2011; Bezaz & Kacha, 2021; Schuldt, 2013), imagery (Di Cicco, Zhao, Wijntjes, Pont, & Schifferstein, 2021; Gil-Pérez, Rebollar, & Lidón, 2020; Schifferstein, Lemke, & de Boer, 2022; Underwood & Klein, 2002), shape (Chitturi, Londoño, & Henriquez, 2021; Koo & Suk, 2016; Raghubir & Greenleaf, 2006) and size (Çakır, Balagtas, Okrent, & Urbina-Ramirez, 2021; Chu, Tang, & Hetherington, 2021; Reister & Leidy, 2020), can guide inferences about specific product attributes (e.g. taste, healthfulness and quality) and further influence choice and consumption. The underlying assumption for such packaging design elements to be able to affect product evaluations and consumer responses is that they become salient to consumers (Bloch, 1995; Orth & Malkewitz, 2008).

DOI: 10.4324/9781003000822-4

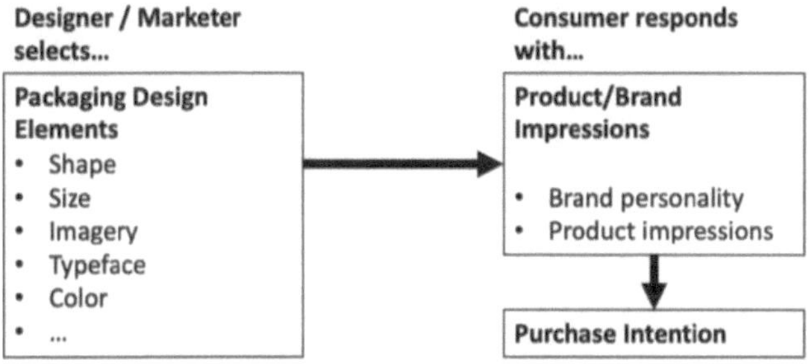

Figure 3.1 Packaging design as communication tool.
Source: Adapted from Orth and Malkewitz (2006).

By contrast, the *holistic* perspective is rooted in Gestalt theories in psychology. According to this approach, consumers' first impression of a package is shaped by its holistic appearance, meaning that the response is derived from the blending of all the elements of the package, in contrast to when these cues are presented in isolation. Although individual cues alone may convey a certain meaning to consumers, the holistic approach supports the view that consumers hold a universal perception and engage in a higher-level general processing first before they process the individual cues of a package (Bloch, 1995, 2011; Orth & Malkewitz, 2008). Next, we discuss these two approaches in greater detail, but mainly emphasise the atomistic perspective since research in this area is largely dominated by the atomistic perspective.

Holistic Perspective of Packaging Design

Researchers have paid attention to the holistic view of the design of packaging for decades by proposing different typologies of holistic designs. Initially, Bloch (1995) proposed a concept based on the physical characteristics (ergonomics, durability and recyclability) of the packaging design but extended it later to a three-dimensional structure of aesthetic, functional and symbolic dimensions (Bloch 2011). Dhar and Wertenbroch (2000) classify packaging design as either hedonic or utilitarian. Similarly, Orth and Malkewitz (2008) propose five packaging archetypes rooted in unique combinations of design factors: massive, contrasting, natural, delicate and nondescript (see Figure 3.2). They find these archetypes to be associated with distinct brand impressions. For example, massive designs are associated with low sophistication and average sincerity; contrasting designs generate high excitement, high ruggedness, low competence and average sincerity; and natural designs convey sincerity and sophistication but only average ruggedness. These results suggest that a brand that aims to build a certain impression should strategically choose the type of archetype

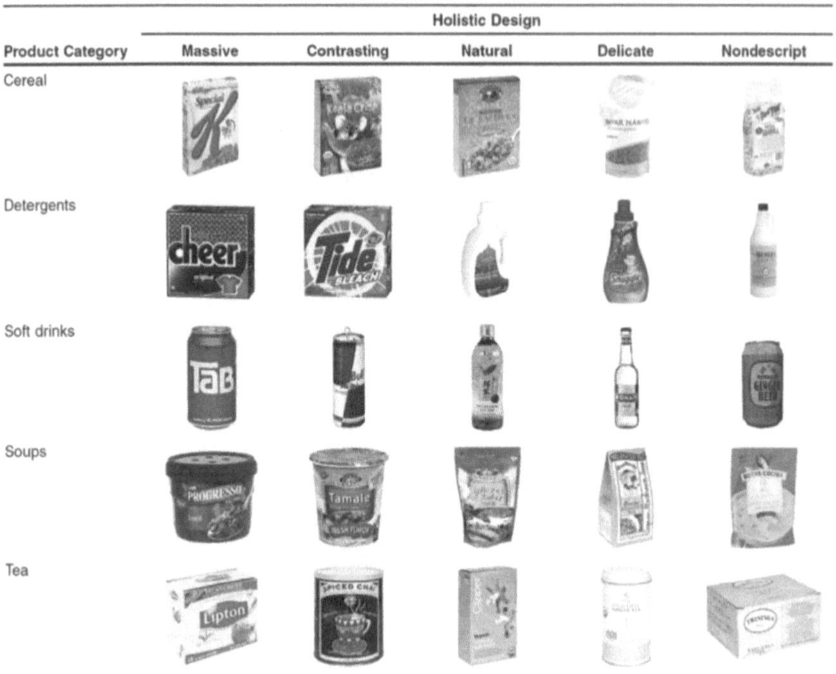

Figure 3.2 Examples of holistic packaging design archetypes across categories.
Source: Orth and Malkewitz (2008).

its packaging should convey. In addition, Orth and Malkewitz suggest that sincere brands should have natural packaging designs, exciting brands should have contrasting designs, competent brands should have delicate designs, sophisticated brands should have natural or delicate designs and rugged brands should have contrasting or massive designs. They further highlight that a holistic packaging design may communicate brand personality and improve brand positioning more effectively.

In proposing five packaging archetypes, Orth and Malkewitz (2008) offer three reasons to support the efficacy of holistic packaging designs. First, drawing from Gestalt psychology (Köhler, 1967) and studies on non-conscious preference formation (Zajonc, 1968), specific design elements are perceived as exhausting a higher-level dimension. Second, some studies have been able to identify and articulate fundamental factors that could explain the differences among different types of designs (e.g. Henderson & Cote, 1998). Third, different packaging designs are able to extricate a variety of impressions (Aaker, 1991; Keller, 1993a). Thus, holistic packaging designs are generalisable to brand impressions.

Atomistic Perspective of Packaging Design

Marketing research on packaging design mainly follows the atomistic approach and considers packaging design as comprising a set of elements or cues, each with the potential to shape how consumers perceive the product via the cue. To elaborate, for example, visual images and brand names often serve as critical components to help convey product selling points and portray the product favourably (Clement et al., 2013; Romaniuk & Sharp, 2015). These cues may contain multiple meanings so that they influence consumer evaluations of the product and packaging (Ampuero & Vila, 2006; Ooijen, 2016; Silayoi & Speece, 2007; Van den Berg-Weitzel & Van de Laar, 2001). Consequently, a considerable number of studies have examined the signalling meaning and the psychological mechanism of packaging elements across different markets (Angka, Hémar-Nicolas, Hapsari, & Olsen, 2020; Bakar, Lee, & Rungie, 2013; Celhay, Cheng, Masson, & Li, 2020; Herbes, Beuthner, & Ramme, 2020; Khan et al., 2015; Limon, Kahle, & Orth, 2009).

A review of the relevant literature indicates various approaches to classifying these individual cues. Silayoi and Speece (2007) classify these as (a) *informational* (e.g. product information and labels) and (b) *visual* (e.g. graphics, colour, imagery, shape and size). Underwood (2003) classifies them as (a) *graphic* (e.g. colour, typography and image) and (b) *structural* (e.g. shape, size and material). Konstantoglou, Folinas, and Fotiadis (2021) classify these as (a) *informational* (e.g. information on nutrition and origin), (b) *operational* (e.g. protection and transportation functions), (c) *physical* (e.g. material) and (d) *visual* (e.g. imagery and colour). Festila and Chrysochou (2018) classify them as (a) *informational* (e.g. labels and claims), (b) *graphic* (e.g. imagery and colour) and (c) *structural* (e.g. shape, size and material). Next, we describe each group of packaging elements following the classification Festila and Chrysochou (2018) propose.

Informational Elements

These are all relevant information sources found on the packaging, such as brand name, manufacturer, price, nutrition facts, usage instructions, ingredients, country of origin and food labels. Informational elements either manifest through written language (i.e. specific claims) or a well-established system of symbols (i.e. labels, logos). Informational elements act as explicit communication tools of brand-related values and assist consumers in making informed choices (Silayoi & Speece, 2007). A big stream of research investigates the role of such elements when communicated through packaging in influencing consumer responses. For example, nutrition information on consumer understanding (Grunert & Wills, 2007), health claims and nutrition information on consumer perceptions of product healthfulness (Franco-Arellano, Vanderlee, Ahmed, Oh, & L'Abbé, 2020; Ikonen et al., 2020; Sielicka-Różyńska, Jerzyk, & Gluza, 2021), organic labels on taste evaluations of products (Lee, Shimizu, Kniffin, & Wansink, 2013) and purchase behaviour (Bauer, Heinrich, & Schäfer, 2012; Marian, Chrysochou, Krystallis, & Thøgersen, 2014).

Graphic Elements

These are the elements printed on the package, which offer visual identity to the food product (e.g. colours, imagery and typeface). These elements manifest through pictorial and stylistic representations and therefore rely on specific metaphors and learned associations to convey meaning (Festila & Chrysochou, 2018; Karnal, Machiels, Orth, & Mai, 2016). *Colour* is an important graphic element and an important source of package differentiation (Ampuero & Vila, 2006). Colour in packaging can attract attention (Klimchuk & Krasovec, 2012), elicit emotion (Liao, Corsi, Chrysochou, & Lockshin, 2015) and convey brand values and associations (Kauppinen-Räisänen & Luomala, 2010). Further, colour can influence judgements about products, such as its taste, caloric content and healthfulness (Becker et al., 2011; de Sousa, Carvalho, & Pereira, 2020; Karnal et al., 2016; Schuldt, 2013). For example, Schuldt (2013) shows that a candy bar was perceived as healthier when the package had a green, rather than a red, calorie label, even if both labels displayed the same calorie content. Becker et al. (2011) find that highly saturated packaging colours lead consumers to anticipate a more intense product taste. Further, Karnal et al. (2016) demonstrate that consumers perceive food products to be healthier when lighter colours are used on the package.

Imagery includes all the illustrative and photographic elements on the package. Imagery is an effective design tool that influences product evaluations and consumer responses (Underwood & Klein, 2002). Moreover, it attracts attention (Bartels, Tillack, & Jordan Lin, 2018) and elicits emotions (Liao et al., 2015). Apart from the presence of imagery on packages, the actual content influences consumer evaluations and even their actual behaviour. Madzharov and Block (2010) show that when more units of products are displayed on the package, consumers believe that the pack contains a higher product quantity and therefore consume more of it. Machiels and Karnal (2016) show that images depicting unprocessed foods misguide consumers into perceiving the product as more natural. Similarly, Smith, Barratt, and Sørensen (2015) report that when photographs of a product are present on the package, the product's taste is perceived as more natural.

Typefaces are any verbal information found on packaging, which is printed in a certain typeface that can be a custom-designed or a standard one. Typefaces are integral to a brand's communication and, in several cases, an asset (Ward, Yang, Romaniuk, & Beal, 2020). They may influence consumer responses and product perceptions (Henderson, Giese, & Cote, 2004; Velasco & Spence, 2019). Henderson et al. (2004) find that natural script fonts are more reassuring and pleasant than simple, non-ornate fonts. Gofman, Moskowitz, Fyrbjork, Moskowitz, and Mets (2009) suggest that fancy fonts on labels could help consumers feel more energised. Velasco, Woods, Hyndman, and Spence (2015) suggest that round typefaces lead to sweeter taste perceptions, whereas Otterbring,

Rolschau, Furrebøe, and Nyhus (2022) find that round typefaces increase consumers' preferences for sweet food.

Structural Elements

These are the elements that give form and texture to the product (e.g. shape, size, material and transparency). Structural elements aim to not only enhance functionality and ease transport and storage, but also improve feel and support usage. In a similar fashion to other packaging design elements, structural elements can support consumers in drawing inferences about a product. A distinguishing feature of structural elements is that in addition to using their visual sense, consumers may use their haptic and motor senses to create inferences by touching or handling the package. This additional feature makes the study of structural elements more complex (Krishna & Morrin, 2008).

Shape is one of the most well-documented structural elements of packaging design. A product's package shape can influence its identity (Underwood, 2003) and consumers' perceptions and preferences regarding a product (Raghubir & Greenleaf, 2006). Shape can further bias consumers in estimating the volume and quantity of products, which may even influence the amount of product they consume. For example, Sevilla and Kahn (2014) demonstrate that incompletely shaped packaging designs are perceived to be smaller. Koo and Suk (2016) show that products in wider or convex packages are perceived to have more calories and as less healthy than products in elongated or concave packages. Last, Fenko, Lotterman, and Galetzka (2016) show that angular shapes make consumers perceive products as healthier.

The *size* of a package can significantly affect quality judgements, attract attention and influence volume perception and consumption (Reister & Leidy, 2020). Moreover, the package size can influence quality judgements (Yan et al., 2014), with smaller packages being rated more favourably than larger packages. The authors attribute this effect to smaller packages being associated with higher unit prices. Thus, larger packages are perceived with smaller unit prices. Another impact of package size is on consumption, with larger packages resulting in perceptual biases about the actual content amount and overconsumption (Chandon, 2013).

Material refers to the types of material (e.g. paperboard, metal, plastic and glass), as well as to their properties (e.g. texture and transparency). The type of material that companies decide to use for their products depends on several factors, such as the product type; transportation conditions; storage, protection and shelf-life requirements; disposal/recycling and cost. The material may further affect the types of inferences consumers make about products. For example, the material is the main heuristic that consumers use to assess whether packaging is sustainable, with plastic being considered a less sustainable material than other types of material (Ketelsen, Janssen, & Hamm, 2020; Nguyen, Parker, Brennan, & Lockrey, 2020). Perceptions about material can also transcend to evaluations about products. Products that are wrapped in more sustainable materials are perceived to be more natural and of higher quality and can even make food products appear healthier to consumers (Magnier & Schoormans, 2015). Another example is that when the packaging texture is rougher, consumers believe the product to be crunchier and harder (Piqueras-Fiszman, Velasco, & Spence, 2012).

Transparency is another characteristic of packaging that influences consumer response to packaged food and their consumption amount. For instance, Deng and Srinivasan (2013) show that transparent packages can influence consumption. When the packaging contains small and visually attractive foods (e.g. candies), consumers eat more from a transparent package than from an opaque package. Conversely, when a transparent package contains larger foods (e.g. cookies), consumers eat less. Further, products in transparent packages are perceived to be not only of higher quality, more attractive and fresher but also healthier than products in opaque packages (Simmonds & Spence, 2017).

Technology as a Packaging Design Element

One packaging element that has not attracted equal attention in the marketing literature is packaging technology. Packaging incorporates a variety of technologies that essentially play an important role in consumer decision-making. These technologies can serve various roles, such as providing additional properties to the product (e.g. extending its shelf life; antimicrobial), supporting information provision (e.g. traceability and spoilage) or even enhancing and supporting the product's use (e.g. ergonomics and recycling).

Packaging technologies in the food domain are generally divided into active and intelligent/smart packaging technologies. *Active packaging* solutions have the ability to interact with the packed food and modify its conditions to improve its safety and extend its shelf life. *Intelligent and smart packaging* solutions have the ability to monitor the condition of foods during storage and transport and give information about the product quality and its shelf life as well as any other relevant product-related information. Another domain of packaging technologies that are currently emerging are those that aim to promote the *sustainability* of food and packaging systems (e.g. biodegradable or recyclable packaging). Such packaging solutions have been gaining the interest of companies that aim to meet their sustainable development goals, as well as of researchers who aim to support their adoption and to understand consumer responses. Indeed, sustainability and ethical consumption is a developing theme among Asian consumer packaged goods (CPG) consumers, given that they are increasingly expressing their intention to buy sustainable products (Lewis, 2016; YouGov, 2021). For instance, Lewis (2016) reveals that the number of such consumers ranged from as high as 79% for Indonesia to more than 50% for Singapore and Hong Kong, and that consumers in China (65%) and India (73%) also indicated similar intentions to support products with sustainability characteristics.

Problems that packaging technologies experience are consumer scepticism and reluctance to adopt these technologies, which often derive from a lack of both

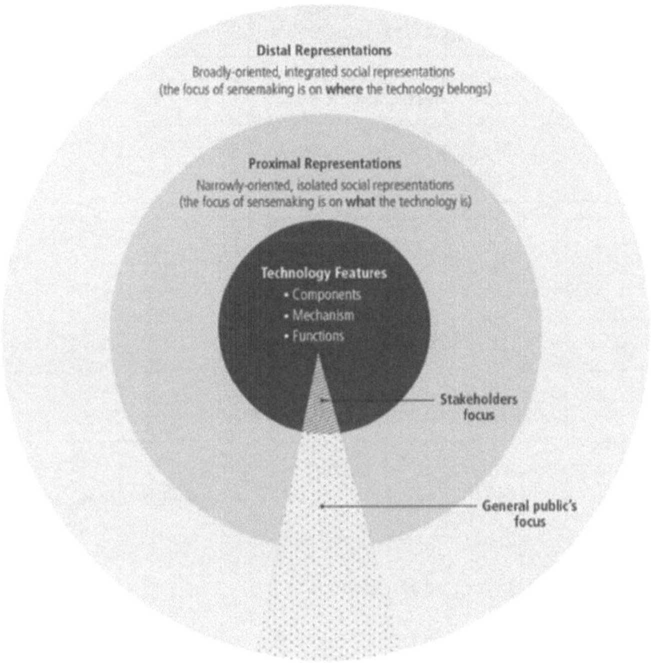

Figure 3.3 Sensemaking framework of active packaging technologies.
Source: Festila et al. (2021).

understanding and of proper communication about their features. In fact, the way consumers understand and assess packaging technologies differs from how professionals (e.g. material scientists and packaging manufacturers) understand them. In the case of active packaging technologies, Festila, Chrysochou, Hieke, and Massri (2021) show that the way consumers make sense of such complex technologies is by focusing on what the technology is (distal representations) and where the technology belongs (proximal representations), as illustrated in Figure 3.3. Thus, whereas professionals emphasise the technology features more, the general public focuses on the proximal and distal representations, which need to be better understood and communicated to ensure acceptance.

Classifying Packaging Design Cues

Figure 3.4 summarises the discussion thus far of different cue types by presenting them as a classification. From the discussion, and aided pictorially by this classification, some interesting observations can be derived about how each of the three types of packaging design elements works and the way that marketing research has studied these types.

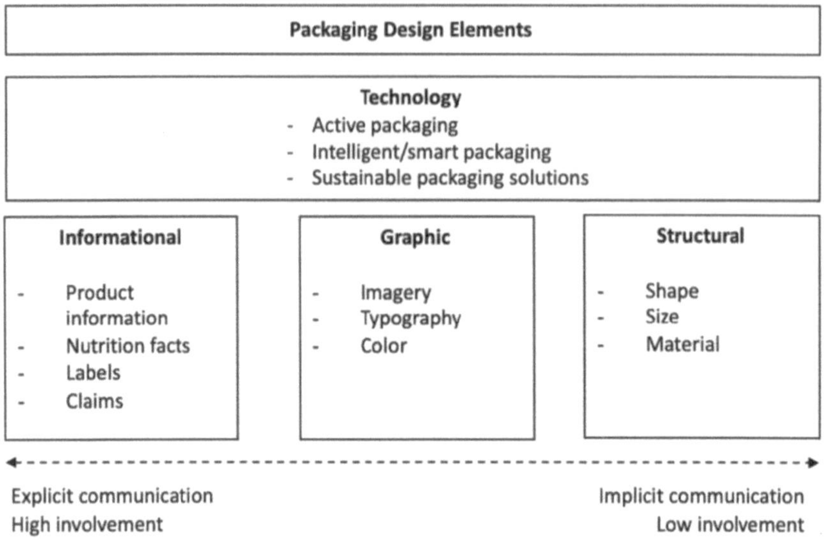

Figure 3.4 Classification of packaging design elements.

First, informational elements can be considered essential in communicating brand values explicitly, whereas graphic and, more importantly, structural elements communicate brand values in a more implicit manner. For example, health and nutrition claims explicitly communicate that a product is healthy, but imagery of exercise conveys health in an implicit manner. Such implicit elements are found to be more powerful in conveying the brand value (Chrysochou & Grunert, 2014). Notably, informational elements are often regulated (e.g. health claims and nutrition facts), whereas graphic and structural are mostly not. It is therefore important for marketers who make decisions about packaging design to realise both the way that each element conveys the desired brand value and the power of graphic and structural elements in communicating such values implicitly. In fact, since an implicit form of communication entails that marketers have less control over how the message is decoded by consumers, it requires to be carefully understood before use.

Second, an important distinction between the groups of packaging design elements is the level of involvement that consumers require to process them. Informational elements require more conscious and elaborate processing, and thus higher involvement, but graphic and structural elements require less processing, and thus lower involvement (Festila & Chrysochou, 2018). Hence, implicit packaging design elements become particularly effective under low-involvement situations, such as during food choice. In these situations, consumers do not allocate sufficient cognitive resources to read the available written information and therefore simplify their decisions by using certain visual heuristics (e.g. colours or images on the package). Such heuristics might attract initial consumer attention and become diagnostic sources of information, creating expectations for the content of the package (Silayoi & Speece, 2007).

Third, the majority of the literature has focused on the impact of informational elements on consumer responses, which form a deliberate way of communication. The shift to the study of graphic and structural elements has attracted less attention and only emerged in the past two decades, especially in the realm of focusing on unconscious processes and their potential to nudge consumers. However, the processes are not yet well understood, which leaves room for additional research that needs to be done in this area.

Fourth, it should be noted that although the atomistic approach focuses on the study of a specific type of element, these elements do not work in isolation and may interact with other elements, thus creating unknown interaction effects (i.e. adding or subtracting from the total effect). In fact, this is a structural weakness of the atomistic perspective, with the holistic perspective having a significant advantage to overcome this weakness. Such interactions are mostly ignored in the literature, and in cases that are considered, their origin is less understood. For example, the impact of a label on the health evaluations of a product may also depend on the colour of the label, with green labels increasing the perceived healthfulness of a product (Schuldt, 2013). Not considering the holistic perspective may also explain the lack of consistency across studies on the impact of specific elements on consumer response. Another particular point to note is that technology forms a layer that is often interconnected with the other three types (e.g. the material type is essentially a technology element).

A Case Example: Organic Food Packaging Designs

Following the description presented in the previous section, packaging elements for a product should not be chosen at random but should be carefully decided to account for the expected effect on consumers' perceptions about the product as well as their choice, consumption and decisions at the product's end life (e.g. recycling). Several companies are aware of these issues, and if such decisions are made, for certain product categories a systematic use of packaging elements should be observed. To answer this question, Chrysochou and Festila (2019) conducted a content analysis of organic food packaging designs and investigated how these designs are constructed by identifying the packaging design elements that are prominent for organic products.

Chrysochou and Festila (2019) analysed informational (verbal) and graphic and structural (visual) packaging design elements (see Figure 3.5). Their main finding is that organic food products use systematically informational elements such as logos and claims to explicitly communicate (and certify) that a product is organic. Packages of organic products differ from their conventional counterparts in graphic and structural elements as well, with organic products having more paper material, white and green colours and images of nature.

Product: Milk
Country: Denmark

<u>**Visual Packaging Design Elements**</u>
Material: Paperboard
Colour: Blue
Imagery: Yes
Nature: Yes

<u>**Organic Product Verbal Packaging Design Elements**</u>
Organic Product: Yes
Presence of Organic Label: Yes
Origin of Organic Label: National and EU
Organic Label Position: National on front of product; EU on side of product
Presence of Organic Claim: Yes

Figure 3.5 Packaging design of organic products.
Source: Chrysochou and Festila (2019).

Discussion

Packaging design elements play a key role in conveying brand values (Bloch, 1995) and are part of a brand's unique identity (Ward et al., 2020). Most studies have followed the atomistic perspective and focus on exploring the impact of specific design elements on various forms of consumer response (e.g. attitudes, perceptions, emotions and choice). As we show in this chapter, an atomistic approach would lead to individual elements falling into one of three categories, namely informational, graphic and structural. This, in turn, begets a question: How do the three categories relate to each other, and do they work in tandem or in a standalone manner in influencing consumers?

According to Sugermeyer (2021), the influence of packaging on consumers occurs over four phases. During the first phase of a shopping journey, packaging serves to first capture shoppers' attention as they approach store shelves. This stage may occur non-consciously as the shoppers are selectively attracted by graphic elements, particularly those that they find intriguing or familiar. The consequence of this behaviour is that shoppers would defocus on the majority of competing products in the category and choose to focus only on a few products.

In the second phase, shoppers would consciously evaluate or compare the noticed products, and here is where informational elements come into play. The informational elements should provide the information to assure shoppers that their requirements are met. Phase three then follows, in which shoppers start to scrutinise a product that attracts their interest, and in this phase, the packaging, both informational and structural, should assure shoppers that they would be making the right choice. Last, once the consumer buys the product, the role of a packaging is to facilitate consumption through suitable structural elements, as well as to help build brand equity by embodying the brand's intrinsic identity, such as through graphic elements, including brand logos or other distinctive brand assets. This last step would make the product easier to recall and be noticed during the first phase in the next shopping trip.

Thus, it can be understood from this description of a shopper's journey that a packaging system is complex and much work is still needed to guide marketers fully on how to make comprehensive use of packaging as a strategy tool. Many parameters need to be considered, which essentially often makes decisions about packaging design a challenging task for managers. An atomistic perspective to study the effect of packaging elements is certainly a way forward, but it is necessary to gain improved understanding about how such elements convey brand values and how consumers process such elements.

In this regard, of particular interest to this book is how the packaging designs of identical products from major Western CPG firms are modified when these products are exported to, and sold in, Asian markets. An ancillary question would thus be how effective these changes are in appealing to local Asian consumers. To answer these questions, the next chapter discusses the marketing strategy of localisation (versus standardisation) as it applies to packaging. The chapter also reports a case study on actions taken by major Western CPG brands to localise their products for the China market.

4 Methods of Measuring Packaging Effectiveness

Packaging Effectiveness

What Is Packaging Effectiveness?

According to Robertson (2005), packaging has four essential functions: containment, protection, convenience and communication. Effective packaging serves all these roles well.

Containment: A primary objective of packaging is to contain the product. Packaging facilitates product transportation, storage and distribution. Structural elements, such as the material, size and shape, play an important role. Thus, effective packaging needs to ensure the proper choice of such elements that enhance its containment objective.

Protection and preservation: Another important function of packaging, especially in the case of foods, is protection and preservation. The key function is to protect the product from environmental factors (e.g. light, oxygen and bacteria) that may affect its shelf life, quality and safety. Simultaneously, packaging may protect the environment from the product, such as from odours and spillage. Similarly, the structural elements of packaging play an important role (e.g. material), and thus, effective packaging ensures that these functions are met to a large extent.

Convenience: Convenience is a packaging function that mainly refers to how end users interact with a product. Thus, it relates to how easy it is for consumers to buy, use and recycle the product. Emphasis regarding this function has been on aspects such as the packaging size, and the ease of using and of opening/closing it. More recently, more emphasis has been put on the need for packaging to support the functions of reuse and recycling, thus addressing calls for improved sustainability (Ketelsen et al., 2020). Effective packaging that addresses convenience needs to address these goals, and apparently, emphasis is again placed on the choice and advancement of structural elements.

Communication: Packaging is often referred to as the 'silent salesman' since it is a medium that allows communicating product-related information and is a brand-building platform. The informational and graphic elements of the package play an important role in communication, but as mentioned in Chapter 3,

DOI: 10.4324/9781003000822-5

structural elements can act as implicit communication tools. Effective packaging is thus one that communicates the desired product values, and, therefore, all elements support this goal.

From a marketing perspective, effective packaging is one that primarily meets the communication function. Indeed, the emphasis put on marketing departments in companies, and the majority of the research that has been conducted, has focused on understanding this aspect. However, all functions may contribute to the brand-building process, and thus, from a marketing perspective, effective packaging needs to communicate these functions successfully to customers. For example, the improved sustainability of packaging or the extended shelf life on introducing a new packaging solution are functions that may still lie in the marketing managers' role for they need to be communicated properly to customers.

Principles of Effective Packaging Design

Consistent with the classic response hierarchy models, consumers pass through cognitive, affective and behavioural stages before committing to a purchase decision (Kotler & Keller, 2016). These stages define the main steps of the communication process when it is regarded as a linear, sequential process. The attention, interest, desire and action (AIDA) model—the best-known model—suggests that consumers go through four distinct stages: attention, interest, desire and action.

According to the AIDA model, packaging acts as a stimulus that supports the communication process at each stage. Therefore, effective packaging helps a product to be noticed on the shelf (attention), supports consumers in their learning about the product (interest), creates a positive response and engages consumers with the product (desire) and commits consumers to purchasing the product (action; Table 4.1).

To address the effectiveness of packaging at each stage, a number of methodological approaches are used and recommended (Table 4.1). For example,

Table 4.1 Packaging role following the AIDA model and examples of methods

Stage	*Packaging role*	*Example of methods used to address the role*
*A*ttention	Attract attention on the shelf	Eye tracking, survey (e.g. aided awareness), virtual reality
*I*nterest	Raise curiosity and obtain information about the product	Survey (e.g. objective knowledge), focus groups
*D*esire	Create positive emotions and engagement with the product	Neurophysiological methods (e.g. EMG), observation
*A*ction	Commit to a purchase	Panel data, ethnography, experiments

eye-tracking studies are often used as a method to assess consumers' level of attention to products on the store shelf. Emotional responses to packaging can be assessed with neuropsychological methods (e.g. electromyography [EMG]). These methods are discussed in the following section.

Methodological Approaches to Assess Packaging Effectiveness

Marketing research that addresses research questions related to packaging effectiveness is rather rich in methods. This section describes the most commonly used ones, followed by research examples that address issues of packaging design effectiveness. More specifically, these methods are grouped into (a) *qualitative methods* (focus groups, in-depth interviews, observation and ethnography), (b) *survey research*, (c) *experimental research*, (d) *panel and sales data*, (e) *neuromarketing methods* (eye tracking, facial EMG and electroencephalography [EEG]/functional magnetic resonance imaging [fMRI]) and (f) *sensory measures.*

Qualitative Methods

Qualitative methods are used to provide insight and an in-depth understanding of the problem considered. This section describes the following commonly used methods: focus groups, in-depth interviews, observations and ethnography.

A *focus group* is a popular qualitative method in which a group of participants discusses a topic of interest (Krueger & Casey, 2014), which allows them to interact with each other. Focus groups are usually used when the objective is to explore consumer views of new forms of packaging and ways to increase consumer acceptance. Participants can be asked about their perceptions, feelings, beliefs and ideas about packaging. As a method, focus groups are preferred when the researcher is interested in obtaining a deeper understanding of the problem studied.

Mitchell, Moodie, Critchlow, and Bauld (2019) conduct a focus group study with adolescents from the United Kingdom to study their awareness of, and responses to, standardised cigarette packaging. They find that standardised packaging is not appealing, which increases the saliency of health warnings. Nguyen et al. (2020) conduct a series of focus groups with consumers to explore their understanding of eco-friendly packaging. A key finding is that consumers use packaging material as a heuristic to assess packaging eco-friendliness, but their knowledge about manufacturing technologies is limited.

An *in-depth interview* involves conducting interviews with a few respondents to explore their perspectives on a topic of interest for the researcher (Boyce & Neale, 2006). The method is appropriate when the researcher wants to gain deeper insights into the topic and produce precise answers. These interviews have several advantages, such as these allow respondents to take their time to provide their views, allowing a less structured approach in collecting answers from respondents, and avoiding peer influence that often is present in focus groups. The method is preferred when it is not easy for the researcher to gather respondents and conduct focus groups (e.g. informants are packaging design experts).

Ford, Trott, and Simms (2016) use in-depth interviews combined with observations to explore how older consumers interact with packaging. Their findings show that older consumers are more vulnerable during packaging interactions. For example, vision changes make it difficult for them to read packaging information in small fonts, while changes in strength and agility make it difficult for them to open jars. Their findings suggest the need for firms to empower older consumers through improved packaging design that reduces vulnerability. With in-depth interviews as part of a multi-step method, Williams, Lindström, Trischler, Wikström, and Rowe (2020) discuss how households interact with packaging to reduce food waste. Among their key findings is that packaging size makes it difficult for households to buy the right amount and minimise food waste.

Observation systematically describes events, behaviours and artefacts in the social setting chosen for study (Marshall & Rossman, 2014) and is commonly used in anthropological fieldwork. Observation involves observing the behaviour of individuals or groups systematically and purposefully to learn about a phenomenon of interest to the researcher. Observation can be considered a data collection method for ethnographic studies and case studies, and while quantitative, can also be more systematic or structured. Nemat, Razzaghi, Bolton, and Rousta (2022) use photo-based observation to explore why consumers miss-sort plastic food packaging and assess how packaging design can facilitate sorting behaviour. Their findings suggest that the structural elements of the packaging design (e.g. packaging form, size, durability and haptic aspects) and the lack of declarative information on the package decrease the perceived value and recyclability. Wells, Farley, and Armstrong (2007) conduct in-store observations to explore shopping behaviour and to understand how consumers evaluate the packaging design of private label brands and how it influences their purchase decisions. Their findings show that a large proportion of consumers are aided by packaging at the point of purchase.

Ethnography is an interpretative research method performed in the native environment to obtain a deeper understanding of people's behaviour. Ethnography offers researchers a naturalistic approach to studying a phenomenon, compared with the controlled conditions, which introduce biases, used in other methods. Ethnographers engage with participant observation while using other sources of data (e.g. archives, questionnaires and in-depth interviews).

The emergence of digital communications and social media has shifted ethnographers to studying online behaviours, which has resulted in an ethnographic research method termed *netnography* (Kozinets, 2019). In packaging design, ethnographic methods are rather rare, but netnography offers an emerging research opportunity to study the effectiveness of packaging. Ilich and Hardey (2020) use netnographic investigation together with in-depth interviews to explore the motivations, intentions and marketing implications of sharing online images of secondary packaging while unboxing products. The authors suggest four motivations behind such practices: identity, documentation, socialisation and aesthetics. Sattlegger (2021) uses an ethnographic approach to understand why single-use packaging remains a constitutive element in self-service supermarkets. The author suggests that packaging offers diverse roles and that acknowledging these roles within the framework of workplace practices can reduce packaging waste.

Survey Research

A common method to address packaging effectiveness is *surveys.* Survey research involves collecting information from a sample of individuals through their responses to questions (Malhotra, 2019). The method is preferred when the purpose of the study is to assess the characteristics (e.g. attitudes, beliefs and perceptions) of a target population and collect their opinions on a phenomenon of interest to the researcher. Surveys explore correlations between constructs. Thus, experimental designs should be preferred when the aim is to establish causality. Surveys can be either cross-sectional (i.e. conducted at a given point in time) or longitudinal (i.e. repeated over time for the same or different samples). Longitudinal surveys allow observing consistency and patterns in the phenomenon studied, and the development of the phenomenon over time.

Hao et al.'s (2019) survey of consumers in China analyses the influential factors of their willingness to pay for sustainable packaging. The authors find that although consumers have insufficient knowledge regarding sustainable packaging, they are willing to pay for it. Moreover, through factor analysis, they find that the key drivers affecting consumers' willingness to pay are the environment, green packaging quality, commodity and packaging price. Monteiro, Silva, Ramos, Campilho, and Fonseca (2019) use a survey to explore how the packaging industry adapts to sustainable development goals. Through a questionnaire administered to packaging manufacturers, the authors find that a small proportion considers eco-design an important issue. Further, eco-design development is correlated with the materials' shelf life, certified suppliers and supplier reputation.

Experimental Research

Experiments are used as a research method for establishing cause-and-effect relationships (Malhotra, 2019). Causality means that one event (independent variable) will cause another event (dependent variable). Experiments can take place online (*survey experiments*), in a laboratory (*laboratory experiments*) and in the field or in-store (*field experiments*). In packaging design research, the main objective of an experiment is to explore whether a change in packaging design can result in changes in a focal measure of interest to the researcher (e.g. purchase behaviour and perceptions).

Yarar, Machiels, and Orth (2019) conduct experiments to explore the impact of food packaging shape on consumers' evaluation of the healthiness of foods. The authors find that consumers perceive that slim packages contain more healthy foods, an effect that is pronounced among women with increased body mass index. Baccarella et al. (2021) demonstrate through a series of experiments how consumption-supportive packaging functionality (i.e. packaging with a function that helps users achieve their consumption goals and that supports the objectives they have in mind when using it) influences consumers' purchase intentions. Their findings show that this packaging functionality leads to higher purchase intentions and that this effect is explained through increased perceived product meaningfulness.

Panel and Sales Data

Panel data are observations of the same individuals over a period. Such data are normally obtained from households (e.g. household panel data) or from retailers (e.g. scanner panel data or sales data). Household panel data are collected from specialised market agencies that recruit households to register their daily purchases, such as GfK (www.gfk.com) and Kantar (www.kantarworldpanel.com). Scanner panel data or sales data are retailer data from either their sales records (sales data) or loyalty programs (i.e. programs that reward their members with offers) that can track their purchase behaviour (scanner panel data). Panel and sales data allow researchers to explore actual customer behaviour, which is more advantageous than the other methods discussed thus far. However, such data are difficult to obtain, and thus, studies, especially those that focus on packaging design effectiveness, are in fact limited.

Kühn, Strelow, and Gallinat (2016) use sales data to observe how the sales of chocolate bar from packaging designs can be predicted from functional brain activation. They find that fMRI data of product designs can be better predictors of sales data than explicit liking judgements. Yang and Raghubir (2005) demonstrate, as part of a multi-method series of studies, that elongated containers are perceived to have less product quantity. They use sales data to validate their hypotheses by showing that the purchased amount of beer is higher in cans than in elongated bottles.

Neuromarketing Methods

The emergence of multisensory marketing has influenced research on packaging design effectiveness (Spence, 2016; Velasco & Spence, 2018), with the emergence of studies using neuromarketing research methods that aim to measure responses from human senses. Neuromarketing methods provide measures of consumer reactions that are more accurate and more objective than self-reported measures (e.g. surveys). However, it is suggested that for studies that assess food packaging design effectiveness via neuromarketing methods, they complement self-reported measures or other forms of behavioural measures since they help explain consumer behaviour better (Moya, García-Madariaga, & Blasco, 2020). Such methods require special equipment and are conducted in controlled settings (e.g. in labs), which can be obstructive for participants and thus introduce response bias. In this section, we discuss the key methods used in this research domain: eye tracking, facial EMG, EEG, fMRI and sensory measures.

Eye tracking involves the study of participants' eye movements to assess their visual attention. Using devices called eye trackers, the visual paths of participants are recorded to assess where they look and how they navigate. In packaging design research, eye tracking can have several applications. For example, it can be used to assess whether consumers notice a product on the shelf, how they navigate when scanning through packaging design and the information on packaging that they notice, which consequently influences their choice.

Husić-Mehmedović, Omeragić, Batagelj, and Kolar (2017) use eye tracking to assess the package features within the (canned) beer category that affect consumers' visual attention. Following a process-based framework, they find that physical (e.g. size) and semantic (e.g. text and imagery) package features affect attention during the 'orientation' phase, whereas attention shifts to the brand during the 'discovery' phase. Similarly, Clement et al. (2013) use eye tracking, both in-store (using mobile eye trackers) and in the lab, to test the influence of packaging design features on visual attention. They find that size and shape affect attention, but not colour and brand-related features. However, they find that consumers' attention is rather fragmented during grocery shopping and is disrupted by the shelf display.

Facial EMG is used to measure emotional reactions and their valence and has been used to assess responses to marketing stimuli, such as packaging design. Facial EMG records facial muscle movements, mainly the activity of the corrugator and the zygomatic major muscle, which is known to be associated with mood state. The method measures the conscious and subconscious expression of emotions, unlike self-reported measures used in surveys.

Liao et al. (2015) use facial EMG to assess emotional responses to food packaging elements by focusing on colour, images and typefaces. Their findings suggest that images generate an emotional response that both self-reported and physiological measures can measure. In contrast, colours and typefaces generate an emotional response that only self-reported measures can measure. Vila-López, Kuster-Boluda, and Alacreu-Crespo (2021) use facial EMG combined with other explicit and implicit measures of emotion to test whether results converge by using the case of packaging design from low-fat products. One of their key findings is that while informative packaging elements (messages) have stronger power to elicit emotions in virtual environments than graphic elements (colour) do, the opposite occurs in physical environments.

EEG and *fMRI* are methods to record brain activity and identify the parts of the brain that are activated when an individual is exposed to certain marketing stimuli, such as packaging. Human brains react subconsciously to external events, and recording these reactions help researchers understand consumer behaviour. Reimann, Zaichkowsky, Neuhaus, Bender, and Weber (2010) use fMRI to assess the psychological properties of packaging design. They find that aesthetic packaging design increases brain activation in the nucleus accumbens and the ventromedial prefrontal cortex. Moreover, they find that fMRI responses are correlated with a paper-and-pencil measure of affective product involvement. Hubert, Hubert, Florack, Linzmajer, and Kenning (2013) use fMRI to measure neural responses to product packages under different impulsive buying tendencies. Their key findings show that attractive packages activate brain regions associated with an impulsive system and reward.

Sensory Measures

Although the visual and emotional aspects of packaging design have dominated research on packaging design effectiveness, the number of studies that examine packaging as a medium offering a multisensory experience is growing (Krishna, Cian, & Aydınoğlu, 2017; Velasco & Spence, 2018). For example, studies have examined how packaging feels in terms of touch (Decré & Cloonan, 2019; Littel & Orth, 2013) or smells (Velasco & Spence, 2018). Decré and Cloonan (2019) find that glossy packaging is perceived as less rough, thinner and lighter than matte packaging. Such haptic perceptions significantly affect perceptions of quality and purchase intentions. Similarly, the presence of visual stimuli on packaging may influence taste perceptions. Van Rompay, Fransen, and Borgelink (2014) find that movement visuals on packaging connoting upward movements result in a less concentrated smell when presented in the top-left region of the package.

Deciding on the Method

The decision on the method to use depends on the actual research question and several factors that the researcher needs to consider, such as the budget, the time and the availability of respondents. Certainly, each method has both advantages and shortcomings (Chrysochou, 2017), and even though they are not discussed in this chapter, they need to be taken into consideration.

A key decision on which method to use relies on the type of consumer response due to exposure to packaging that the researcher is interested in measuring. As presented in a flowchart in Figure 4.1, if researchers attempt to measure opinions and aim to obtain a deep understanding of the phenomenon, they could use methods such as focus groups and in-depth interviews. Instead, if their objective is to obtain an overview and not to explore the phenomenon in depth, they should prefer survey methods. If the aim is to measure behavioural responses that can be assessed in their natural environment, then the appropriate methods are observations, ethnography and panel and scanner data. Experiments, neuromarketing methods and sensory measures could be used if the behavioural responses cannot be assessed in their natural environment.

Cross-Cultural Considerations in Measuring Packaging Effectiveness

This section presents a final note regarding the generalisability of findings related to packaging effectiveness across cultures and the consistency in results from the methods used. Although research results can be assumed to be generalisable to

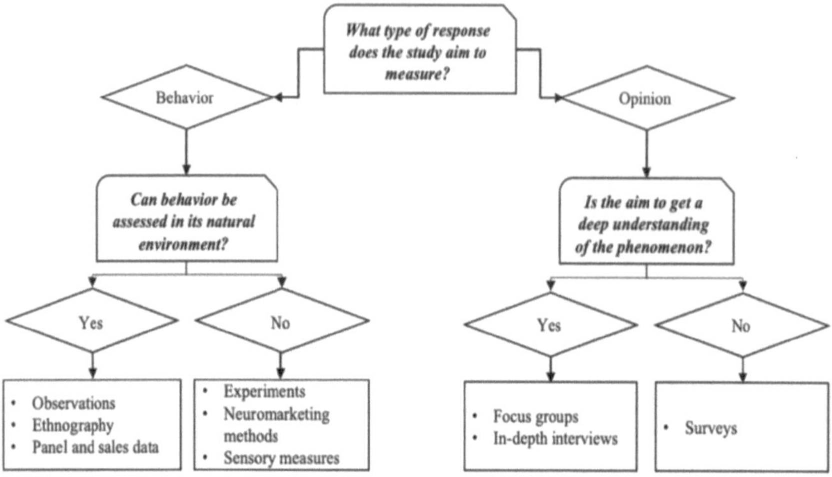

Figure 4.1 Flowchart in choosing a research method.

humankind, this is not always the case (Cheon, Melani, & Hong, 2020; Henrich, Heine, & Norenzayan, 2010b). One reason is that a substantial proportion of research in social sciences has been produced with participants from Western, educated, industrialised, rich and democratic societies, which represent 12% of the world's population (Henrich, Heine, & Norenzayan, 2010a). The lack of generalisability could also be the result of the method used. For example, in survey research, Likert scales may result in differences owing to the different ways in which they are assessed across cultures (Heine, Lehman, Peng, & Greenholtz, 2002), a phenomenon commonly observed in mixed-worded scales (i.e. scales that contain both positive- and reverse-worded items; Wong, Rindfleisch, & Burroughs, 2003). In choosing the method and tools (e.g. questions, scales and equipment), it is critical that researchers consider such possible shortcomings.

In addition, the role of culture should be taken into account, given the calls for increased understanding of its impact on packaging design (Machiels & Orth, 2019). There is growing evidence that the symbolic meanings of design elements (e.g. colour, typeface and shape) are culturally dependent (Machiels & Orth, 2019). Thus, despite the emergence of global consumer cultures that call for package standardisation, there are calls for strategies that embrace package localisation (Celhay et al., 2020). Similarly, research on packaging design effectiveness should embrace a localised strategy by adjusting the methods and tools used. Further, they should avoid generalised assumptions and instead focus on measuring responses in the actual markets in which consumers will interact with packaging.

Thus far in Part I, we have discussed role of packaging beyond its role as a physical container to it being an effective marketing communications tool. We also introduce the concept of packaging elements and importantly show that packaging elements may shape consumers' perceptions of a product via its packaging and, consequently, their purchase behaviour. Part I further highlights the differences between a holistic versus atomistic perspective of packaging designs and describes the common methods used by marketers to measure packaging effectiveness when selecting or modifying packaging design from a cross-cultural approach perspective. Drawing on the concepts of standardisation and localisation, the next chapter concludes Part I by setting the theoretical foundation to explain why localising the packaging of foreign products is effective for Asian markets.

5 Packaging Localisation

A Case Study of Supermarkets in China

Huda Khan and Wayne Wang

Packaging Localisation as a Marketing Strategy

In an era of rapidly increasing internationalisation, international firms face the challenge of deciding whether to standardise or localise their business strategy when they sell their products in foreign markets (Busnaina et al., 2021; Cleveland & Chang, 2009; Romaniuk & Sharp, 2015). This ongoing discussion has appeared in various business domains for decades, including new market entry strategies, business operations, management and international advertising (Bamiatzi, Bozos, Cavusgil, & Hult, 2016; Hill & Hult, 2016; Yip & Hult, 2012). Standardisation of international marketing strategies refers to using a uniform marketing mix (price, product, distribution and promotion) to yield similar strategies across countries (Chiou & Droge, 2015; Kravets & Sandikci, 2014; Sheth, 2011). Localisation of marketing strategies refers to tailoring marketing communication approaches according to the differences in the mentioned aspects in the target market where products are sold (Bai et al., 2021; Khan & Lee, 2020a; Rigby & Vishwanath, 2006; Vrontis, 2005).

Studies that advocate standardisation approaches argue for the developing trend of globalisation, integration and homogenisation in the global market (Melewar & Vemmervik, 2004; Tan & Sousa, 2013). They assert that cultural differences across global markets, both physiologically and psychologically, are lessening, and marketers are better off addressing areas of communality rather than tapping into trivial differences among consumers. Owing to the development of ubiquitous technologies and globalisation, international firms stand to maximise their market share and profitability by standardising their marketing strategies and communications (Chiou & Droge, 2015; Yip & Hult, 2012; Zucchella & Magnani, 2016). Consistent with this postulation, a meta-analysis of 152 studies found that executing standardised or localised marketing strategies caused no statistical differences in influences. For instance, the marketing elements (e.g. advertising) in both approaches showed similar functions and strengths that could satisfy consumers' needs across different markets (Cheon, Cho, & Sutherland, 2007). P&G is one of the most successful international companies that executes a standardised business approach. It has focused on eliminating marketing differences between countries and developing standardised

DOI: 10.4324/9781003000822-6

new products since 2005. Using this strategy, the company achieved strong growth and saved about US$800 million in cost annually (Hill & Hult, 2016; Neff, 2003).

In contrast, by drawing out differences in factors such as culture (Cleveland & Chang, 2009; De Mooij, 2010; Hofstede, 2001), religion (Bakar et al., 2021), economy (Hoskisson et al., 2000) or industry structure (Paliwoda et al., 2009; Vrontis, 2005), localisation (also called adaptation) is a critical factor for achieving a competitive advantage over rivals in targeted foreign markets. Owing to the heterogeneity of these factors across markets globally, marketers need to tailor marketing communication strategies according to culture, environment and the industry structure of the country in which they sell their products (Busnaina et al., 2021; Samli, Wilkinson, & Mutscheller, 2015). Since consumers in different countries may perceive the same brands or products differently, international marketers need to be cognisant of, and cater to, these differences in marketing their products.

Researchers have proffered reasons for adopting a localised approach. That is, according to the cultural consequences theory of Hofstede (2001), people tend to develop 'mental programs' such as behaviour while growing up. These mental programmes are heavily influenced by the individual's surrounding culture. Over time, these programmes convey strong values that subsequently influence their consumption behaviour. Social identity theory (Tajfel, 1978; Turner, 1982) may also explain the need for localisation. Formally, social identity is 'that *part* of an individual's self-concept which derives from his knowledge of his membership of a social group (or groups) together with the value and emotional significance attached to that membership' (Tajfel, 1978, p. 63, emphasis original). Turner (1982, p. 19) further explains that social identity is an 'enduring, multi-faceted system which is carried about in the head from situation to situation', and 'in any given situation a different part or combination of parts of the self-concept could be at work'. In other words, localisation enables consumers to identity with the products, and hence, they will be more biased or predisposed to liking the products and buying them. Consequently, people perform consumption behaviours that they identify with and are culturally unique to them. Indeed, some researchers have contended that cultural differences are so vast and unique globally that standardisation has become unfeasible (Melewar & Vemmervik, 2004; Middleton, 2015).

For example, Velasco, Salgado-Montejo, Marmolejo-Ramos, and Spence (2014) find that participants from different cultural backgrounds respond differently to colour, for the colours convey different meanings and aesthetic values across cultures. Khan et al. (2015) reveal that Pakistani consumers prefer localised verbal cues (in Pakistani) on the packaging of utilitarian products more than they do the standard verbal cues (in English), whereas consumer preference did not differ on the packaging of hedonic products. Similarly, Bakar et al. (2013) suggest that attaching a crescent-moon symbol on the packaging may favourably bias the consumers' perceptions of the product in Islamic markets. Hence, packaging design should follow such cultural trends when targeting Islamic markets in order to maximise marketing efficacy (Bakar et al., 2021; Valentine & Gordon, 2000).

In further support of this proposition, Wan et al. (2014), through a cross-cultural study, find that participants with different cultural backgrounds (from the United States and China) attribute different meanings to colour, therefore behaving differently towards coloured beverages. In the context of advertising, Neelankavil, Mummalaneni, and Sessions (1995) provide evidence that Asian consumers hold more positive attitudes towards Western brands when seeing foreign models in advertising rather than local models, whereas Toffoli and Laroche (2002) reveal that Chinese consumers show a higher preference for an advertisement in a foreign language to that in the native language. Similarly, Beck, Chapman, and Palmatier (2015) show that customising the product and service offering to local demands can increase the value of the product and service in the local market (Herbig, 1998; Massa & Testa, 2009). Consequently, international firms often adapt their marketing strategies, products and services in this way to achieve scale economies, learning effects and local demands (Hill & Hult, 2016).

Further, the same applies to product packaging. Because product packaging plays a key role in marketing communications between consumers and brands (Creusen, Gemser, & Candi, 2018; Krishna et al., 2017; Samiee, 2019; Spence et al., 2019), it is understandable that global brands seek to compete for the attention and favour of consumers in foreign markets by localising their product packaging (Ampuero & Vila, 2006; Khan & Lee, 2020a; Khan et al., 2015; Khan, Lockshin, et al., 2017; see Figure 5.1 as an example). Simply put, adapting the packaging would appeal to the cultural differences and best serve the targeted local markets (Khan et al., 2015; Limon et al., 2009; Van den Berg-Weitzel & Van de Laar, 2001). By adapting or catering to the differences, packaging localisation can tap the uniqueness of each market in order to differentiate and gain advantages against competing brands. Consistent with this argument, research involving 500 UK-based companies revealed that companies adopted localisation

Figure 5.1 Localised packaging of Western CPG brands in non-Western markets.

strategies for reasons such as culture, competition and differences in sociological conditions (Vrontis, 2005).

Another reason for packaging localisation is to make information processing easier and less effortful for consumers. Fitzsimons et al. (2002) contend that verbal cues induce efficient cognitive processing, which lead to effective consumer responses towards product evaluations. This contention resonates with research in marketing communication that shows that local language cues in advertisements are effective because consumers notice, and respond to, relevant native cues present in the elements of communication (Brumbaugh, 2002; Khan & Lee, 2020a; Koslow, Shamdasani, & Touchstone, 1994; Torres & Luna-Nevarez, 2012). In their study on the ease of processing of language cues to facilitate consumers' recall and recognition of the advertised brand, Ahn and Ferle (2008) consider English (foreign) versus Hangul (local) language effectiveness (in terms of recognition and recall) on South Korean consumers when language is manipulated for brand name and body copy of the advertising message. They find that the brand name written in the foreign language with the body copy in the local language is an effective strategy to target South Korean consumers.

Further, others argue that because most consumer purchase decisions occur in stores spontaneously at store shelves, rather than being premeditated (Chandon et al., 2009; Kacen et al., 2012; Stilley, Inman, & Wakefield, 2010), brands are not purchased unless they are noticed at the point of purchase. When little thought goes into product purchasing, point-of-purchase communication determines brand choice decisions (Bell, Corsten, & Knox, 2011; Clement et al., 2013; Kacen et al., 2012). Consequently, packaging localisation makes it easier for consumers to recognise brands, thereby shaping their decision at the point of purchase.

Packaging Elements in Localisation Practices

As discussed in Chapter 3, apart from being a container, a product packaging contains elements or cues to assist consumers in evaluating the product (Bartsch, Riefler, & Diamantopoulos, 2016; Ng, Chaya, & Hort, 2013). Packaging elements not only help consumers to make informed choices (Delgado-Ballester & Fernandez Sabiote, 2015; Wood, 2007) but may also deliver different meanings to them (Khan et al., 2015; Limon et al., 2009). Chapter 3 categorises packaging elements as information, graphic or structural, possibly overlaid by technology. Following from this classification, we now review studies on how packaging cues within each of these three categories are often localised.

Informational Elements

Product labels, a classical and most common type of informational element, typically state the ingredients or usage instructions. They support an efficient cognitive process to offer product information to consumers regarding product

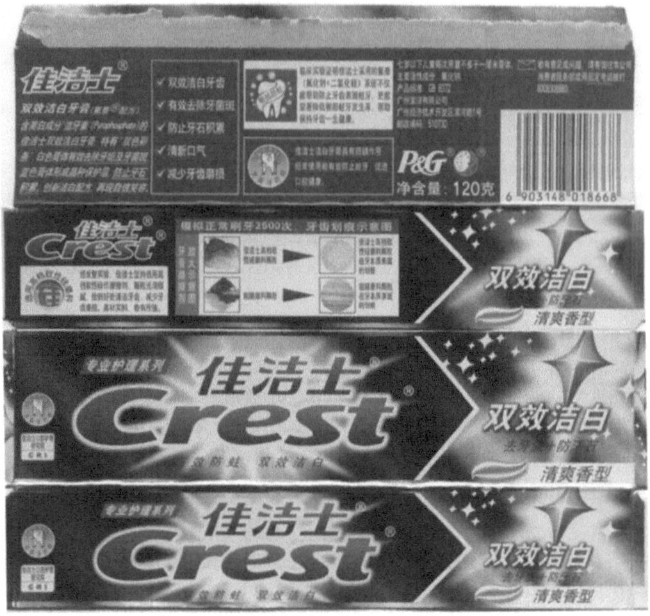

Figure 5.2 Localised product labels.
Note: 'Crest' by Sinosplice is licensed under CC BY-NC 2.0.

characteristics (Marsh & Bugusu, 2007; Mueller & Szolnoki, 2010; Souiden, Abdelaziz, & Fauconnier, 2013). In packaging localisation practices, labels have similar effects when they are translated into local languages. Dublish, Seaton, and Laskey (2015) highlight that using a local language, rather than a foreign language, on the packaging may facilitate local consumers to gain product knowledge more rapidly. Consider the case of China. Western brands often localise their marketing communication messages simply because of a lack of English literacy amongst Chinese consumers (Tai, 1997). An example of packaging that has informational elements is provided in Figure 5.2.

Localising labels also has an important role in ensuring regulatory compliance. In the context of China again, the 'China Food Safety Law and Packaging regulation' mandates that Western brands sold in China must clearly display in *Chinese*, information such as the product ingredients, the company/manufacturer details, the manufacturing date and shelf life, the packaging size (net volume/content) and even the production licence for some products.

Graphic Elements

Visual images may also convey product characteristics to consumers (Ampuero & Vila, 2006; Kahn, Deng, & Krishna, 2010). Serving as surrogates to inform consumers about the product content, such images (e.g. a 'cooked pasta' image) are often associated with product features (e.g. look, taste and smell) that increase cognitive elaboration and improve consumer evaluation (Clement, 2007; Madzharov & Block, 2010). However, in addition to providing product information, visual images may contain non-product-related meanings and associate with people's feelings, such as their sense of belonging, personal image and social identity (Ooijen, 2016; Van den Berg-Weitzel & Van de Laar, 2001). For example, national flags (as visual images) may cause a range of consumer self-identity (e.g. patriotism) recognition (Gelb, 2002). Similarly, images such as those of bamboos and dragons, which have a strong cultural identity in China, could affect Chinese consumers' choice making (Yetts, 1912). For example, see the dragon-themed Coca-Cola can in Figure 5.3. Further, using Chinese models as an image on food packaging can deliver a better effect to appeal to Chinese consumers (He & Wang, 2017). Consumers may perceive the 'love theme' between family members from the image and associate it with Chinese culture and social custom.

As regards the visual elements, Wood (2007) further contends that some *symbols or icons*, (e.g. the 'Gluten Free' icon) can simplify product judgement during the purchasing process by delivering relevant information on product benefits to consumers in a more straightforward manner. Meanwhile, other symbols may disseminate cultural meanings to consumers. For example, the picture of a crescent moon on packaging could connect with Muslim consumers' sense of religious or cultural identity, and favourably bias their perceptions of the product (Bakar et al., 2013, 2021). See the examples of symbols in Figure 5.4.

Colour is another graphic element that can carry cultural meanings although the meaning may vary across cultures (Aslam, 2006; Jacobs, Keown, Worthley, &

Figure 5.3 Visual images and symbols on the packaging.

Note: 'Coca-Cola 2000 Cambodia Chinese new year' by Jose Roitberg is licensed under CC BY-NC-ND 2.0.

Figure 5.4 Gluten-free and crescent-moon symbols.

Figure 5.5 Coca-Cola launched 'Chinese New Year' theme.

Note: 'Coca-Cola does Chinese New Year' by Christopher is licensed under CC BY-NC 2.0.

Ghymn, 1991; Velasco, Salgado-Montejo, et al., 2014). For example, Ampuero and Vila (2006) reveal that yellow has a special meaning in Spain. Spanish consumers often relate yellow to patriotism, which thus appeals to their ethnocentric tendencies to buy locally produced products. Similarly, green symbolises the love theme in Japan. Among the Chinese, red symbolises luck, joy and happiness, and Coca-Cola launched a 'Happy Chinese New Year' theme during this festive period in China, as exemplified by Figure 5.5. These examples illustrate that international marketers need to be cognisant of the use of appropriate colours when designing product packaging for different markets.

Figure 5.6 Graphic packaging elements with distinctiveness.

Note: '8-bit Nike Logo' by The Logo Smith and 'Old Apple Logo Web 2.0' by Alistair Israel are licensed under CC BY-NC-ND 2.0.

Brand-related images, such as *brand logos* or *brand names*, are another type of graphic element. Examples include the 'Just do it' tagline of Nike, the cartoon character on M&M's chocolate and Coca-Cola's red sinuously curved bottle (see Figure 5.6 for examples). The rationale underpinning the efficacy of brand elements is that they influence consumer behaviour in various ways. First, they appear on packaging for both the store shelf and advertising. Second, they deliver focal points to attract visual attention and thus make the packaging stand out on crowded retail shelves. Third, they affect consumers' memory not only on the spot but also at home. Last, they help buyers to recall and describe the brand more easily, thereby influencing their purchase decisions (Lynch, 1991; Romaniuk & Sharp, 2015). Further, the more consistent the packaging cues built into consumer memory are, the more they influence consumers' choice making (Romaniuk & Nenycz-Thiel, 2013; Romaniuk & Sharp, 2015).

In particular, in a supermarket environment, consumers tend to spend little time shopping for consumer packaged goods (CPG) (Pieters & Warlop, 1999; Sharp, 2010). For example, Dickson and Sawyer (1990) assert that almost 50% of supermarket shoppers spend less than five seconds at the point of purchase. In line with this finding, Keller (1998) suggests that building brand awareness through distinctive brand images would facilitate consumer choice making by ensuring brands are easily recognised and remembered by buyers (Garber, 1995; Romaniuk & Sharp, 2015; Sharp, 2010).

Taken together, this discussion suggests that if brand elements are consistently used, they may leave a strong impression in the consumer's mind. Consistent brand cues in packaging may communicate with consumers more efficiently and ease their decision-making because they perform little evaluation at a shelf and hence use these cues as shopping shorthand (Romaniuk & Nenycz-Thiel, 2013; Sharp, 2010; Warlop, Ratneshwar, & Van Osselaer, 2005; Zhang, 2014). This is especially true for brand elements that are unique and are proprietary properties of the firms owning the brands. Thus, it appears that brands may not want or need to localise such brand elements, for doing so would result in them losing their identity and hence competitiveness.

Figure 5.7 Localisation of brand elements.

Note: '7 Up New Look (Take 2)' by Julien G. and 'Coca-Cola Chinese design' by renaissancechambara are licensed under CC BY 2.0.

That said, examples abound of Western brands localising brand elements for Asian markets. For example, consider the two products in Figure 5.7—one has the traditional Western brand logo, and the other is localised. 7 Up is localised by using the red dot together with the character xi (喜) to embody happiness and luck in the Chinese culture. The logos of Coca-Cola are shown in Chinese and English. Likewise, Dove chocolate added a tagline in Chinese to say 'Eating Dove brings good luck every year' on its gift-wrapped chocolate, which has cultural significance in traditional Chinese culture (He & Wang, 2017; Zhou & Hui, 2003).

Structural Elements

Packaging literature also indicates that the structural packaging elements, including *packaging size, shape and materials*, may deliver product information (Folkes & Matta, 2004; Scott, Nowlis, Mandel, & Morales, 2008; Silayoi & Speece, 2007). For example, *packaging materials*, such as glass, allow consumers to examine the product content, and thereby communicate the intended product benefits (e.g. flavour and smell; Marsh & Bugusu, 2007), attract their visual attention (Peck & Childers, 2006), boost their willingness to touch the packaging (Peck & Childers, 2006) and increase their purchasing propensity (Liu, Batra, & Wang, 2017). Furthermore, packaging materials could also have a particular influence on consumer behaviour because the material can invoke affective responses, such as pleasure and excitement (Chen, Barnes, Childs, Henson, & Shao, 2009; Tu, Yang, & Ma, 2015). In China, it is for this reason that gift packs often come with silk ribbons during the Chinese New Year to symbolise fortune and joy (Zhou & Hui, 2003).

Pack size is another structural element localised for some Asian markets. For example, driven by affordability, many CPGs in India, particularly in the rural regions, are sold in smaller packs, or 'low unit packs' in industry terminology (Bose, 2019). Another example is the milk packs sold in China, which are smaller than those sold in Western markets apparently because of lactose intolerance among the Chinese population in general (daxueconsulting, 2021).

Having discussed how packaging elements—in the three dimensions of informational, graphic and structural—are being localised for Asian markets, we now turn our attention to a case study that compares how major Western CPG brands differ in their packaging for China versus a representative Western market, Australia.

A Case Study of Packaging Localisation of Consumer Packaged Goods in China

To understand how Western CPG brands localise their packaging in Asia, it would be useful to have external validation of real-world events. Therefore, we conducted a case study involving an extensive audit of packaging localisation by Western CPG brands in China's supermarkets and compared the difference between the localised packaging for the China market from the packaging for the same product in a typical Western market—Australia.

We selected two food (chocolate and soft drinks) and two non-food (shampoo and cleaning detergent) categories for both countries. For Australia, we collected data from the two largest supermarket chains (Coles and Woolworths) in the city of Adelaide. For each brand (e.g. a can of Coca-Cola), the collected data regarding the product packaging serve as the *standard packaging* of the product, as would be typically found in Western markets. As a non-Western country, China's culture and language differs markedly from those of Western countries, thereby making the country suitable for this study. The strong consumption landscape in China has also ensured that major Western CPG brands are readily available in major Chinese supermarkets. For China, we collected data in two foreign-owned supermarkets (Walmart and Carrefour) and two local supermarkets (Lianhua and CR Vanguard) in the city of Shanghai. For each brand, we examined both popular local and well-established Western brands that were available on the shelves in the identified supermarkets. This represented the *localised packaging* of the product, tailored for a non-Western market.

Data Collection Procedure

We collected data directly in situ via a videography technique to capture granulated details of packaging designs (Cooper & Schindler, 2006). Direct observation has been widely used in marketing to capture product appearance via mechanical devices (Parasuraman, Grewal, & Krishnan, 2006). The principal advantage of direct observation is that it can capture minute details of the packaging designs. We visited the supermarkets with an inconspicuous eyewear camera and *exhaustively* captured on video all products in the studied categories on the supermarket shelves.

For example, for Coca-Cola, three product variants (Classic, Zero and Diet) with four packaging sizes and types (aluminium can, and large, small and mini plastic bottles) were recorded as they appeared on the shelves. For each product, each face (front, back, top and bottom) of the packaging was first video-recorded and later screen-shot to produce static pictures. In total, 145 packaging stills on 34 brands were obtained in a Western market (Australia) and 240 packaging stills on 41 brands in a non-Western market (China) across four major CPG categories. Table 5.1 summarises the captured categories and brands.

Table 5.1 Captured brands and packaging

Brand	*Category*	*Product*	*Packaging type*
Coca-Cola	Soft drink	Classic/Zero/Diet	Can/Large, small and mini bottle
Sprite	Soft drink	Lemon/Icy lemon/Zero	Can/Large, small and mini bottle
Fanta	Soft drink	Orange/Apple/Peach	Can/Large, small and mini bottle
Pepsi	Soft drink	Classic/Diet	Can/Large and small bottle
7 Up	Soft drink	Lemon	Can/Large and small bottle
Mountain Dew	Soft drink	Lemon	Can/Large and small bottle
Sunkist	Soft drink	Orange	Small bottle
Kinder	Chocolate	Egg/Bar	n/a
Dove	Chocolate	Milk/Dark/Fruit	Bar/Pouch/Gift packed
M&M's	Chocolate	Original/Peanut	Can/Pouch/Gift packed
Hershey's	Chocolate	Cookie/Dark/Milk	Bar/Pouch
Hershey's Kisses	Chocolate	Cookie/Dark/Milk	Pouch
Reese's	Chocolate	Peanut milk	Cup
Snickers	Chocolate	Peanut milk	Bar/Pouch/Value pack
Cote Dor	Chocolate	Milk	Box
Pantene	Shampoo	Original/Flavoured/Man	Bottle (large and small)
Head & Shoulders	Shampoo	Original/Flavoured/Man	Bottle (large and small)
Dove	Shampoo	Original/Flavoured/Man	Bottle (large and small)
L'Oréal	Shampoo	Original/Flavoured	Bottle (large and small)
Clear	Shampoo	Original/Flavoured/Man	Bottle (large and small)
Schwarzkopf	Shampoo	Original/Flavoured	Bottle (small)
OMO	Detergent	Original/Flavoured	Small and large bottle
Attack	Detergent	Original/Flavoured	Small and large bottle
Dettol	Detergent	Original/Flavoured	Small and large bottle
Mr Muscle	Detergent	Drain/Bleach	Bottle
Morning Fresh	Detergent	Flavoured	Small bottle
Comfort	Detergent	Flavoured	Bottle
Purity	Detergent	Original	Small bottle

Data Coding and Analysis

We used an abductive reasoning approach to generate insights into the intended meanings of the captured packaging elements (Glaser, 2008; Walton, 2004). Abductive reasoning is appropriate for this study because it provides a suitable data processing procedure (data fishing and theoretical sampling) to facilitate the grouping of packaging elements with similar meanings. By focusing on the likeliest or the best possible explanation for the observed fact, abductive reasoning is able to identify the explanation most preferred over the others in the inference procedure (Timmermans & Tavory, 2012). Abductive analysis shares criteria with grounded theory, such as focusing on theoretical sampling, requiring theoretical sensitivity and developing new theories. However, the logical form of abductive reasoning is distinct from grounded theory. In detail, grounded theory is developed based on the inductive method and lets the theories emerge inductively rather than applying any analytical strategy. Researchers criticise the 'induction focus' as the key weakness of grounded theory. Abduction, instead, building on grounded theory, aims to analyse qualitative data for theoretical construction and innovation. This approach rests on the systematic analysis of existing theories and evidence and tends to reveal fresh empirical findings to foster theoretical innovation (Kelle, 2010; Locke, 2007). Abduction also logically differs from the other types of reasoning, such as induction and deduction. Inductive reasoning starts with observing specific objectives in a limited scope and tends to generate conclusions that are likely rather than certain. The results found by adopting this approach are operative in the scope but lack universal generalisation.

The abductive reasoning approach comprises three steps: *coding, comparison* and *inference*. Briefly, the *coding stage* involves examining each captured video or picture of a packaging and determining the elements on the packaging that have been localised and the type of elements (e.g. product label and colour) that are being localised. We approached three experienced international marketing managers from different CPG corporations to volunteer as independent coders (see their profile in Table 5.2). Given their extensive experience with CPG brands and packaging strategies, these managers have the knowledge and expertise to act as coders. All three coders were bilingual (English and Chinese) and familiar with both Australian and Chinese culture. In the coding stage, each coder worked independently to determine the purpose of each packaging element. Next, the three coders attempted to group the packaging elements into subcategories depending on the intended meaning of the packaging elements. Then, we requested them to provide keywords and a descriptive title for each subcategory. In the reasoning procedure, they were required to consider the following questions as criteria:

1 Does the given title best describe the meaning of the element?
2 Does the given title clearly fit the design intention?
3 Does the given title best explain or interpret anything of significance to consumers?

Table 5.2 Coders' profiles

Coders	*Gender*	*Location*	*Business activity*	*Management level*	*Tenure (years)*	*Native language*	*Foreign language skills*	*Coding language*
1	M	Australia	Sales & Marketing	Senior Director	15	English	Mandarin Cantonese	English Mandarin
2	M	Australia	Marketing	Director	12	Mandarin	English	English Mandarin
3	M	Australia	Product Marketing	Manager	10	Mandarin	English Spanish	English Mandarin

Although the coding took place individually, the second *comparison stage* brought the three coders together to compare their individual results for consistency and to resolve any differences. This stage also allowed each coder to *revisit* the captured data and return to the same observation more carefully, especially when the other coders obtained different coding outcomes. This process allows the coders to modify their earlier results, check for consistency and discover more suitable explanations with generalisability for the packaging elements. For example, the container colour (white) of OMO (a laundry detergent brand) was coded as a physical element (the title of a subcategory) by one coder, whereas it was identified as a unique brand element (the other title of a subcategory) by the others. A three-step process resolved these disagreements as follows. First, the coders reviewed the literature regarding the disputed element. Second, they visited the brand websites (Australian and Chinese) to understand the brand history. Last, they explained the rationale behind their classification decisions. The coders together with the author then attempted to come to a consensus. Disagreements in coding were found in 2% of the Australian and 3% of the Chinese data, and these were discarded. Since only a few classifications were discarded, these exclusions were not expected to affect the findings.

In the third and final *inference stage*, the coders collectively determined whether and how related subcategories could be grouped together to exhaust a higher level dimension. To arrive at the dimensions, the keywords generated by the coders in the first stage (coding) to describe the subcategories were compared and analysed. A packaging element may fall into more than one dimension. For example, colour may be both symbolic and distinctive; red represents auspicious aspects in many Asian cultures, whereas a particular shade of red is synonymous with Coca-Cola. Apart from three elements (the shape, colour and material of the physical container), all other elements were classified into one dimension only.

Results

Using the three-step abductive reasoning process, we were able to classify each localised element into the three categories discussed in Chapter 3 and the earlier part of this chapter. Before we discuss the findings within each dimension, we first summarise the overall results. Table 5.3 shows the elements and dimensions we found to have been localised for the China market. Table 5.4 shows the proportion of products that practised the particular localisation.

Informational Dimension

For the first dimension, the keywords included cognition, product feature, product information, quality and product content. These packaging elements provide information about product attributes and support consumers' cognitive evaluation of the product (Marsh & Bugusu, 2007; Mueller & Szolnoki, 2010). As Table 5.4 shows, across the four product categories, 100% of the elements under the informational dimension were localised. In particular, product

Table 5.3 Localised packaging elements within the three dimensions

Attributes	*Informational elements*
Product-related attributes	- Product labels (ingredients, instructions) - Product Name - Functional symbols - Visual images
Company-related attributes	- Company information - Manufacturer information
	Graphic elements
Culture-related attributes	- Colour - Symbolic symbols - Visual images
Non-culture-related attributes	- Symbolic symbols - Visual images
Brand-related attributes	- Brand name - Tagline - Brand logo - Unique pattern or visual image
	Structural elements
Physical distinct attributes	- Size

Table 5.4 Localised proportion of the dimensions

Product category	*Informational (%)*		*Graphic (%)*				*Structural (%)*
	Product informationz	*Company information*	*Culture related*	*Non-culture related*	*Brand verbal*	*Brand visual*	*Physical*
Soft drink	100	100	9	27	91	9	18
Chocolate	100	100	15	5	70	10	0
Shampoo	100	100	0	0	100	8	15
Cleaning detergent	100	100	0	36	93	0	36
Mean (food)	100	100	12	16	81	10	9
Mean (non-food)	100	100	0	18	97	4	26

information and company information categories were translated into Chinese for *all* brands. This finding concurs with those of studies that suggest that the language used in marketing communications plays a significant role in facilitating consumer cognitive thinking (Dublish et al., 2015; Toffoli & Laroche, 2002).

The literature abounds with studies that show consumers use information to cognitively evaluate products and make purchase decisions (Hong & Wyer, 1989;

Tybout, Calder, & Sternthal, 1981). Information processing theory (Bettman, 1979) postulates consumer behaviour as a mechanistic consequence of how people think, process incoming information and subsequently retrieve object-relevant thoughts from memory. The theory rests on the idea that human beings process the information that they receive, rather than merely responding to it as if such information were stimuli. In the case of packaging, information such as product labels and manufacturer details provide consumers with incoming information against which they compare relevant information they retrieve from their long-term memory, if available. This comparison would lead them to make an informed judgement regarding the product, thereby leading to a buy-or-reject decision.

Graphic Dimension

For the second dimension, the keywords included culture, personal value, personal belonging, social value, brand names, unique logos, visual images, symbols and individual identity. Rather than conveying functional information or benefits to consumers, these packaging elements can invoke affective responses by signalling certain symbolic associations, such as a sense of belonging, personal image and social identity (Ooijen, 2016; Van den Berg-Weitzel & Van de Laar, 2001). For example, graphic elements, such as the picture of a crescent moon (Bakar et al., 2013), or even colour (Aslam, 2006; Velasco, Wan, et al., 2014), can favourably bias product perceptions.

Compared with the informational elements, only 12%, on average, of the graphic elements were localised. That is, 9% of the soft drink brands and 15% of the chocolate brands had localised or newly added cultural elements, whereas shampoo and cleaning detergent brands had none. Overall, the low proportion of localisation with graphic elements implies that Western brands generally see less need to localise these elements, although multiple researchers have emphasised the efficacy of such graphic elements in marketing communication (Bakar et al., 2013; Samiee, 2019; Underwood, 2003).

The efficacy of the graphic dimension may be explained by the symbolic theory prosed by Cooley (1902), which posits the ways in which individuals capture symbolic meanings from others around them and utilise the meanings to shape their self-image and identity. Symbolic theory was then extended from social psychology to the field of marketing and consumer behaviour in the early 1980s. Specifically, Solomon (1983) is the first to reveal that symbolic consumption could explain how consumers' shopping motivations are affected by the symbolic value of products when they associate the symbolic values with their self-identities (also see Dittmar, 1992). Consequently, to take advantage of symbolic consumption in foreign markets, brand managers often attach certain symbolic cues physically on products to create symbolic value to achieve consumer demands (Ekinci, Sirakaya-Turk, & Preciado, 2013). When applied to packaging design, symbolic value refers to certain psychological meanings attached to a product packaging. To influence consumer preference directly, brand managers often attach certain cues on the packaging that convey symbolic meanings (Bakar et al., 2013; Olsen, Slotegraaf, & Chandukala, 2014). For example, Asian consumers prefer brand names in foreign languages for the foreign version may symbolise a higher individual position (Neelankavil et al., 1995).

Packaging elements can also cue consumer memory recall of brands simply because these are visually or semantically distinctive, thus giving the recalled brands a choice advantage (Romaniuk, 2018; Sharp, 2010). These packaging elements, such as unique brand logos, images and patterns, do not carry any functional or symbolic meaning. Examples of these distinctive elements include the cartoon character in M&M's chocolate and Coca-Cola's red sinuously curved bottle. The localisation within the graphic dimension is almost exclusively through translating the brand name (for 91% of soft drinks, 70% of chocolates, 100% of shampoos and 93% of cleaning detergents). Although brands translate the brand name, no brand chooses to change their logo with localisation modifications. One likely reason is the importance of maintaining a universally consistent distinctive image (e.g. logo) regardless of the market in order to aid memory retention and brand recall (Sharp, 2010; Warlop et al., 2005).

Accessibility-diagnosticity theory (Feldman & Lynch, 1988) may explain the influence of such elements within the graphic dimension. Specifically, accessibility refers to the distinctiveness and awareness of a brand and product cues that may cost less effortful cue retrieval from consumer memory, whereas diagnosticity refers to the sufficiency of cues for consumers in assessing product attributes (Fazio, 1990; Unni & Pillai, 2010). Some researchers further suggest that in the product evaluation process, diagnosticity enhances cue exclusivity and distinctiveness (Feldman & Lynch, 1988; Lynch, 1991). Research also reveals that before making choice decisions, consumers tend to first retrieve the most accessible cues from their memory rather than search all possible information. If the cues in their memory are not distinct enough, consumers may rely on their consumption needs and personal values to process brand or product cues (Meyvis & Janiszewski, 2004; Raggio, Leone, & Black, 2014). Such unique brand-related elements are likely to leave a strong impression in consumer memory, thereby influencing their decision-making.

Structural Dimension

Within this dimension, an interesting finding concerns the physical size of the container. We found that most of the Western brands (82% of soft drinks, 70% of chocolates, 85% of shampoos and 71% of cleaning detergents) reduced their packaging size for the China market. This reduction in volume or size may be due to the lower per capita income of Chinese consumers. Another likely reason is that they simply consume these Western brands less, and hence, the smaller package sizes are more appropriate. As mentioned earlier, another plausible reason for the smaller pack size is affordability. Although those in urban areas are relatively affluent, many Chinese in rural regions have lower income levels, and hence the need for smaller pack sizes.

Discussion

In this chapter, first, we offer an academic perspective of how packaging elements are localised and present theoretical support for the efficacies of this localisation strategy in appealing to Asian consumers. Then, we report a case study to externally validate the ground reality by comparing how Western CPG brands localise their products for China across three distinct dimensions—informational, graphic and structural.

However, thus far, we have assumed that packaging elements would fall exclusively into one of three dimensions. It is plausible that a packaging element, depending on how consumers perceive and interpret it, may fall into more than one dimension. For example, colour may be both symbolic and functional; red represents things that are auspicious in many Asian cultures, whereas red font may be used to ensure that consumers pay attention to the words. Similarly, some packaging materials (e.g. silk) can carry both functional and symbolic values due to the cultural influence. The shape and contour of a classic Coca-Cola bottle is not only structural but also serves as a graphic element to facilitate easy recognition by consumers.

As we shall discuss further in Chapter 8, language is another example of a packaging element that may belong to two dimensions. As an informational element on labels, language facilitates comprehension and the cognitive evaluation of a product (Marsh & Bugusu, 2007; Mueller & Szolnoki, 2010; Souiden et al., 2013). By contrast, the mere display of a language, even when the language is not understood (e.g. see Gerritsen & Nickerson, 2010), can sway consumer behaviour (Khan & Lee, 2020a; Salciuviene, Ghauri, Salomea Streder, & De Mattos, 2010). By acting as a graphic cue, language can symbolise cultural values, foster emotional attachment and allow consumers to associate with, and favour, products (Argo, Dahl, & Morales, 2006; Argo & White, 2012; Bishop, Brocato, & Vijayalakshmi, 2015).

Setting aside this overlapping issue of packaging dimension for now—which we will discuss further in Chapter 8—we have thus far provided an academic account of packaging localisation. We have also reported a case study to externally validate what Western CPG brands do to their products in China. It would be interesting to triangulate these two perspectives by determining Western CPG brand manager's views about packaging localisation. This is the primary purpose of the next chapter, in which we report the findings of a series of in-depth interviews with five managers who manage their respective brands in China and five other managers who similarly manage their respective brands in Pakistan.

Part II

Effects of Packaging Localisation in Asian Markets

6 Effects of Product Type and Brand Familiarity in Packaging Localisation

Role of Product Type in Packaging Localisation

Hedonic consumption is those facets of consumer behaviour that relate to the multisensory, fantasy and emotive aspects of product usage experience (Hirschman & Holbrook, 1982). For example, hedonic products (e.g. chocolate and perfume) generate affective responses and experiences in consumers, and hedonic consumption is underpinned by emotional sensations, such as deriving fun from consumption. By contrast, consumers fulfil functional needs from utilitarian products (e.g. laundry detergent) based on the functions performed by a product (Voss, Spangenberg, & Grohmann, 2003). In this sense, the evaluation and purchase of utilitarian products are often accompanied by rationalising that may even involve logically evaluating trade-offs among the attributes of competing products. The different consumption situations imply that product types play a moderating role in marketing communication strategies (Drolet, Williams, & Lau-Gesk, 2007; Dugan, Clarkson, & Beck, 2021; Garrido-Morgado, González-Benito, Martos-Partal, & Campo, 2021).

Underwood et al. (2001) argue that the effects of customised packaging may not be equally strong across product types because of the differential salience of the packaging to consumers. Particularly with hedonic products, prominent and attention-grabbing cues are important (Garrido-Morgado et al., 2021; Van Rompay, De Vries, Bontekoe, & Tanja-Dijkstra, 2012). For example, Leclerc, Schmitt, and Dubé (1994) show that compared with brand names in English, brand names in French as a foreign language were evaluated more positively for hedonic products, but the effects were weak for utilitarian products. This is because when consumers perceive the image of hedonic products as ordinary, their attitudes and preferences decline (Zhan & He, 2012). In another example, Xu and Chen (2016) show that Western cues work better for a hedonic product (lipstick) more than a utilitarian one (lip balm). They reason that hedonic products are more affect driven, and therefore, consumers pay more attention to symbolic meanings attached to cues.

The use of foreign languages in brand names can also enhance perceived foreignness, thereby increasing the hedonic value of products. Steenkamp et al. (2003) argue that, if foreignness is used as a brand attribute in overseas markets,

DOI: 10.4324/9781003000822-8

it would have a strong effect on the purchase intentions of consumers because of the high-quality perceptions associated with these brands. This argument is based on the rationale that suggests the use of an affective symbolic appeal for hedonic products and a rational appeal for utilitarian products (Drolet et al., 2007). It supports the contention that using foreign language cues on product packaging for hedonic products may be particularly effective for non-Western consumers. In contrast, for products that are more utilitarian in nature, aesthetics or intrinsic cues become less favourable (Hagtvedt & Patrick, 2014). Notably, many of the studies arguing the value of symbolic cues and aesthetics for hedonic products are conducted in the advertising domain. Therefore, it becomes crucial to understand whether these findings sustain for packaging communications because packaging is an entirely different medium of marketing compared with traditional advertising.

Role of Brand Familiarity in Packaging Localisation

Consumers' processing of marketing communication may also vary, depending on their familiarity with the brand (Ma, Wang, & Da, 2021; Morgan, Fajardo, & Townsend, 2021). A brand is often classified as strong or weak, depending upon the consumer familiarity level (Aaker, 2012). Phau and Suntornnond (2006) define brand familiarity as consumers' knowledge about a brand that subsists in the particular product category. This knowledge is often based on how brands are compared on their product attributes and which brand acquires unique attributes in comparison to others in the same category (Baker, Hunt, & Scribner, 2002). Consequently, more familiar brands often enjoy higher levels of liking among consumers, which, hence, results in bias in consumers' choice of brands (Rindfleisch & Inman, 1998). For example, the Mars chocolate brand M&M is a strong global brand. As opposed to strong brands, weak brands have been operationalised as those that hold *low level of familiarity* among consumers and *low market share* (Low & Lamb Jr, 2000; Roehm & Brady, 2007). Using this definition, Wonka is a weak brand because consumers are generally less familiar with it than with Mars as a brand of chocolate.

Research suggests that there are differential effects for the packaging strategies of strong versus weak brands (Jo, Nakamoto, & Nelson, 2003; Mueller & Lockshin, 2008). Although the packaging literature offers support for effective packaging design for weak brands in order to develop brand equity, it also provides evidence that the effectiveness of packaging is more crucial for strong brands. The discrepancies in these findings are mainly due to brand familiarity reasons. For example, packaging cues (a picture with illustration of a product) works effectively (in terms of receiving shoppers' attention) for low-familiarity brands as opposed to brands holding higher familiarity (Underwood, Klein, & Burke, 2001). Underwood et al. (2001) further explained that when brands are unfamiliar and consumers do not have much information about intrinsic product attributes (e.g. taste of a product), they draw inferences from extrinsic attributes (packaging cues). This finding shows that packaging cues are more important when brand familiarity is weak.

In contrast to these findings that cues are more important with low-familiarity brands, the power of a brand is argued to reside in consumers' mind (Keller, 1993b), and therefore, packaging cues serve as a tool to communicate meanings and sustain equity for strong brands (Underwood, 2003). Support for this view is provided by the finding that more than 70% of consumers base their purchase decisions for strong brands on product packaging (Wells, Farley, & Armstrong, 2007). The role of packaging effectiveness for strong brands is further heightened in terms of better taste perception when it is wrapped in familiar packaging as opposed to when the same food is provided in unfamiliar packaging (Robinson, Borzekowski, Matheson, & Kraemer, 2007). Another study finds that the choice of a strong brand (wine) is primarily driven by extrinsic attributes (packaging; Mueller, Lockshin, & Louviere, 2010). Furthermore, in-store attention is also important for strong brands, particularly when the consumer is a frequent user of the product (Chandon et al., 2009). Attention can be gauged through using appropriate textual elements (language) on product packaging (Otterbring, Shams, Wästlund, & Gustafsson, 2013).

Early research similarly confirms that consumers utilise different schemas to evaluate strong and weak brands in terms of familiarity (Park & Lessig, 1981). Brand familiarity reduces the cognitive efforts involved in evaluating those brands owing to the established knowledge structures associated with them (Alba & Hutchinson, 1987; Peracchio & Tybout, 1996). Therefore, strong brands gains better cognitive correlations with brand perceptions, processing (Cheung, Chan, & Sze, 2010), product perceptions (McClure et al., 2004), brand recall (Esch et al., 2012) and brand judgement (Yoon, Gutchess, Feinberg, & Polk, 2006). Consistently, familiar brands generally attain better recall and tendency to influence consumer preference (Kent & Allen, 1994; Park & Lessig, 1981). Hence, strong brands evoke reference points in the memory for better product evaluation (Alba & Hutchinson, 1987; Bettman & Sujan, 1987; Rao & Monroe, 1988).

Studies on standardisation and localisation have also shown diverging results for the effects of brand familiarity on communication messages. For example, Esch, Langner, Schmitt, and Geus (2006) find brand familiarity to be insufficient for determining marketing communication effectiveness. They argue that brand managers must develop tactical initiatives to ensure that ensure consumers continue to hold a unique association with the brand and sustain their feeling of contentment with it. Hence, regardless of brand familiarity, managers need to employ standard (e.g. global) and consistent elements in the marketing mix (including product packaging), and enticing experiential cues with the brand so that consumers may think, feel, sense and relate these cues for better brand evaluation (Schmitt, 1999). A brand communication study suggests brand managers should use consistent and standard messaging for familiar brands to stimulate positive consumer evaluations and standard messaging for weak brands to create an association with, and awareness of, the brand (Delgado-Ballester, Navarro, &

Sicilia, 2012). This finding implies that regardless of the brand familiarity, foreign marketers must use a Western language for its packaging.

A Study on the Pakistani Consumer Packaged Goods Market

To provide evidenced-based findings on the packaging type (standard or localised packaging) preferred for product types (hedonic versus utilitarian) with varying levels of brand familiarity (strong versus weak brands), we conducted a study using the Pakistani consumer packaged goods (CPG) market as the context.

Method

Prior to executing the main survey, we conducted a pretest in two steps to identify suitable hedonic and utilitarian categories, as well as suitable brands, for the main survey. To maximise external validity, the main survey needed to include only well-known foreign brands commonly available in major supermarkets across Pakistan. The first step involved visiting 15 supermarkets in metropolitan areas across Pakistan, specifically Lahore and Islamabad. Following the visits, we identified a list of hedonic and utilitarian products, as well as a commonly available major foreign brand for each product. Examples of these products/brands include nail colour/L'Oréal, soft drink/Coca-Cola and shampoo/Pantene. Secondary market data and statistics on market share also confirmed that the identified brands are major players in their respective categories in Pakistan. For example, P&G and L'Oréal are the top-selling brands in the cosmetic and personal care product categories (Euromonitor, 2015). To further generalise the findings, we also ensured that there was a good mix of food and non-food products in the list.

In the second step, we recruited 52 Lahore-based participants via convenience sampling. They rated the 17 products on a 1–7 semantic differential scale anchored on utilitarian and hedonic values (a lower rating means more hedonic value, and a higher rating denotes more utilitarian value). Brand familiarity was also measured using a single item measure, 'Rate your level of familiarity with the brand', adapted from Thompson, Hamilton, and Rust (2005), with values ranging from 1, which denotes *unfamiliar*, to 7, which denotes *very familiar*. The choice of weak brands was based on a supermarket scan to ensure that these brands are either less available or not sold for their respective product category. Using the pretest findings, we ultimately selected eight brands for the study as categorised into different product types and level of brand familiarity (strong or weak; see Table 6.1).

The main survey was conducted using convenience sampling outside the popular shopping precinct in Lahore district. To reduce respondent fatigue and minimise learning effects, each participant only had to answer a questionnaire containing two of the four categories; exhaustively, there were six possible sets of questionnaires (nail colour and shampoo, nail colour and tomato sauce, chocolate and nail colour, chocolate and shampoo, chocolate and tomato sauce, and tomato sauce and shampoo).

Using a within-subject experiment design, we randomly assigned participants to one of these six groups to assess their packaging preference (in standardised

Table 6.1 Products selected for study

Product type	*Product category*	
	Hedonic	*Utilitarian*
Food	Chocolate Strong brand = *M&M's* Weak brand = *Wonka*	Tomato sauce Strong brand = *Maggi* Weak brand = *McCormick*
Non-food	Nail colour Strong brand = *L'Oréal* Weak brand = *Sally Hansen*	Shampoo Strong brand = *Pantene* Weak brand = *Natures*

[English language] versus localised [Urdu language] forms). The final sample contained 371 participants, with roughly 62 participants in each group; 41.2% were males and 58.8% were females. The age ranged from 20 to 50 years and above, with a mean age of 35 and a median age of 36. Roughly 35% of the respondents belonged to the 20–30 years age group, 34% to the 31–40 years age group, 29% to the 41–50 group and the remaining 1% to the above 50 group.

For each set of questionnaires, respondents were asked to rate the packaging likeability and brand likeability, of the two respective product categories to which they were assigned, using a 7-point scale anchored on *dislike* and *like*. Both the standard and the localised packaging of the brand were analysed as within-subjects to determine the differences in liking for the same individual.

Results

First, we ensured that selected brands represented their respective level of brand familiarity or strength. *t*-Test analyses confirmed this manipulation, as shown in Table 6.2.

Next, we analysed the package likeability across the different products and packaging (see Table 6.3). For hedonic products, participants liked the Western packaging more than the localised packaging. This outcome held for all hedonic brands, including food and non-food, M&M's and Wonka (chocolate) and L'Oréal and Sally Hansen (nail colour). For utilitarian products, there were no significant differences in ratings between standard and localised packaging likeability for both strong and weak brands.

Beyond the study's findings that standard packaging is effective for both strong and weak hedonic brands, there are additional noteworthy findings. Familiar brands were more likeable in the standard packaging (English), and more than weak brands in the standard packaging (see Table 6.4).

Another noteworthy finding is the strong and positive correlation between brand familiarity and standard packaging (see Table 6.5). Localised packaging shows significant and negative correlation. Therefore, it can be concluded that localising the packaging of foreign brands tends to reduce the established brand equity in the minds of consumers.

Table 6.2 Paired *t*-test statistics on brand familiarity for strong and weak brands

Hedonic (n = 186)	*Familiarity*	*Product category*	*Mean*	*Mean difference*	*S.D*	*S.E*	*t-Value (Sig.)*
Chocolate							
M&M's	Strong	Food	5.34	2.40	1.48	0.10	22.05 (<0.01)
Wonka	Weak	Food	2.94				
Nail colour							
L'Oréal	Strong	Non-food	5.37	2.21	1.69	0.12	17.82 (<0.01)
Sally Hansen	Weak	Non-food	3.16				
Utilitarian (n = 185)	*Familiarity*	*Product category*	*Mean*	*Mean difference*	*S.D*	*S.E*	*t-Value (sig.)*
Tomato sauce							
Maggi	Strong	Food	5.28	2.44	1.58	0.11	21.07 (<0.01)
McCormick	Weak	Food	2.83				
Shampoo							
Pantene	Strong	Non-food	5.34	2.25	1.67	0.12	18.32 (<0.01)
Natures	Weak	Non-food	3.08				

Table 6.3 Paired *t*-test statistics for standard and localised packaging

Hedonic brands (n = 186)	*Brand tested*	*Mean standard packaging*	*Mean localised packaging*	*Mean difference*	*S.D*	*S.E*	*Sig.*
Chocolate							
M&M's	Strong	5.65	4.96	0.68	1.98	0.14	<0.01
Wonka	Weak	5.17	3.68	1.48	2.36	0.17	<0.01
Nail colour							
L'Oréal	Strong	5.53	4.87	0.66	2.50	0.18	<0.01
Sally Hansen	Weak	4.93	3.75	1.17	2.69	0.19	<0.01
Utilitarian brands (n = 185)	*Brand tested*	*Mean standard packaging*	*Mean localised packaging*	*Mean difference*	*S.D*	*S.E*	*Sig.*
Tomato sauce							
Maggi	Strong	5.36	5.34	0.02	2.10	0.15	0.88
McCormick	Weak	5.66	5.42	0.23	1.94	0.14	0.09
Shampoo							
Pantene	Strong	5.49	5.45	0.03	2.17	0.16	0.84
Natures	Weak	5.51	5.46	0.04	2.45	0.18	0.78

Table 6.4 Paired *t*-test statistics on brand likeability for strong and weak brands

Hedonic brands (n = 186)	*Brand tested*	*Product category*	*Mean*	*Mean difference*	*S.D*	*S.E*	*Sig.*
M&M's chocolate	Strong	Food	5.25	1.19	1.38	0.10	<0.01
Wonka chocolate	Weak	Food	4.05				
L'Oréal nail colour	Strong	Non-food	5.16	1.18	1.48	0.10	<0.01
Sally Hansen nail colour	Weak	Non-food	3.97				
Utilitarian brands (n = 185)	*Brand tested*	*Product category*	*Mean*	*Mean difference*	*S.D*	*S.E*	*Sig.*
Maggi tomato sauce	Strong	Food	5.16	0.89	1.20	0.08	<0.01
McCormick tomato sauce	Weak	Food	4.26				
Pantene shampoo	Strong	Non-food	5.12	1.00	1.28	0.09	<0.01
Natures shampoo	Weak	Non-food	4.12				

Table 6.5 Correlation of strong brands with packaging likeability

Hedonic brands	*Brand tested*	*Product category*	*Standard packaging likeability, r (sig.)*	*Localised packaging likeability, r (sig.)*
M&M's chocolate	Strong	Food	0.54** (<0.01)	-0.03 (0.62)
L'Oréal nail colour	Strong	Non-food	0.61** (<0.01)	-0.23** (<0.01)
Utilitarian brands	*Brand tested*	*Product category*	*Standard packaging likeability, r (sig.)*	*Localised packaging likeability, r (sig.)*
Maggi tomato sauce	Strong	Food	0.43** (<0.01)	-0.05 (0.42)
Pantene shampoo	Strong	Non-food	0.42** (<0.01)	-0.12 (0.08)

Notes

** Correlation is significant at the 0.01 level (two-tailed).

Discussion

Overall, the findings suggest that consumers favour hedonic products with standard packaging, whereas there is no difference for utilitarian products. This finding is consistent for food and non-food products, and for both strong and weak brands. It implies that indiscriminately localising the packaging as Western CPG brands enter Asian markets may not be the most effective strategy. Although some localisation may be necessary for legislative compliance (e.g. non-Western markets often require food labels to be translated to the native language), this study suggests that hedonic products do not stand to benefit, and may even suffer, from localisation, regardless of brand familiarity. For utilitarian products, the type of packaging does not matter.

Some additional findings are also noteworthy. First, foreignness becomes even more important when the brand is strong and hedonic. Second, the brand likeability of strong brands is greater than that of weak brands. Hence, strong brands result in better product evaluation, as similarly found by other studies (Bettman & Sujan, 1987; Kirmani & Baumgartner, 2000; Rao & Monroe, 1988). Third, the confirmation for brand familiarity results also shows that consumers are more familiar with the selected strong brands than with the weak brands. Interestingly, brand familiarity is negatively correlated with localised packaging, which seems to support that localisation may affect the established brand equity of foreign brands in consumers' mind. Indeed, actual industry practice, as highlighted in Chapter 5, is not fully aligned with these research findings. Western CPG brand managers appear to indiscriminately localise their products, regardless of the product type. This finding does call for further empirical evidence regarding standardisation versus localisation packaging strategies. To this end, the next few chapters attempt to shed more light on this issue by considering other marketing factors known to influence consumer behaviour.

7 Additional Effects of Price in Packaging Localisation[1]

Role of Price and Packaging

Price is an important driver of brand choice decisions (Chen, Choubey, & Singh, 2021; Eastman, Iyer, Eastman, Gordon-Wilson, & Modi, 2021). Consumers are often willing to pay differentially (higher or lower) for the same product with different extrinsic attributes, and one such different attribute is distinctive packaging (Steenkamp, Van Heerde, & Geyskens, 2010). Packaging communicates certain values and develops quality perceptions. A brand can attract price premiums when consumers are willing to pay more for it than for similar competitive brands in the market (Aaker, 1996; Rahman, Langner, & Temme, 2021). The price premium is often a strong determinant of market share and brand performance (Ailawadi, Lehmann, & Neslin, 2003).

In brand management literature, the relationship between price and packaging is often associated with symbolic perception theory. This theory postulates that consumers draw certain meanings from environmental stimuli. In this context, the role of perceived quality from the extrinsic cues of the packaging of a brand has been emphasised (Kotler & Keller, 2016). Furthermore, in discussing the willingness to pay a higher price for a brand than for competing brands, scholars assert that brand uniqueness or distinctiveness is critical for consumers' evaluation (Anselmsson, Bondesson, & Johansson, 2014). Marketers often use cues to distinguish their brands from competing brands, and often locate these cues on the product packaging to make it more appealing or to signal quality (Piñero et al., 2010).

Marketing managers can set different perceptions of brands through packaging. Thus, packaging is critical to branding and perception development when the product is marketed across borders (Breetz, 2014). Studies show a strong relationship between packaging and consumers' willingness to pay a higher price. This is because consumers evaluate message information provided on the package to justify the price charged (Breetz, 2014). A study reported that 86% of the respondents were willing to pay a higher price for a nicer package (Silayoi & Speece, 2007).

In Chapter 6, we have already shown the effects of product types and brand familiarity. Now, to understand how price further compounds the interaction of

DOI: 10.4324/9781003000822-9

product types and brand familiarity, we replicate the Pakistani study in Chapter 6 by conducting a study in China with the same product categories.

Method

We use the same product categories and brands as the ones in Chapter 6. Using an online panel provider in China, and similarly to the procedure in the Pakistani study in Chapter 6, participants were assigned to one of six groups with each participant responding to an online questionnaire regarding two product categories. The final sample had 383 participants; 36% were males and 64% were females. The age ranged from 20 to 50 years and above, with a mean age of 35 and a median age of 31. The participants were divided across the six questionnaires, with each questionnaire having roughly 63 respondents.

The online questionnaire comprised two parts that were first developed in English and then translated into Chinese following the standard double-translation procedure (Craig & Douglas, 2005). One part consisted of the same attitudinal questions as in the previous chapter. Participants rated, on a scale of 1–7, how much they liked the focal brand and its standard packaging. Then, they answered the same two questions for the localised packaging.

The second part was a discrete choice experiment (DCE) that captured participants' choice behaviour. Three attributes were considered: packaging, brand and price. The packaging attribute had two levels, where one represented the standard (English) and the other represented the localised (Chinese) packaging, similarly to the examples in Figure 1.1. The brand attribute had three levels. Since the main aim of the DCE was to determine the part-worth utility of the focal brand identified in the pretest, we introduced the two other levels as alternatives to the focal brand in each of the four categories. The first alternative was a brand that was unavailable in the 15 supermarkets across the four cities that we had visited; henceforth, in this paper, we term this alternative as 'secondary brand'. Because this study is not about competition among brands, the purpose of using this unavailable brand is to ensure that consumers concentrated on the focal brands—and their different packaging—in the DCE. The second alternative was named 'Any other brand' to represent a brand other than the focal brand or the secondary brand. Last, as common with the DCE procedure, we included a 'None of these brands' option to represent that participants had no intention to choose any brand.

The price attribute had four levels. To set the price level for each of the focal and secondary brands, we first noted the average market price of each brand during supermarket visits. Then, we set the four price levels at -20%, -10%, + 10% and +20% relative to the average market price. On average, we found that the market price for the secondary brands were about 20% less than those for the corresponding focal brands; this difference is unsurprising for the focal brands are well-established global brands. Last, we set the price for 'Any other brand' as the simple average of the market prices of focal and secondary brand. A summary of the three attributes and their levels are displayed in Table 7.1.

Table 7.1 Attributes and levels for discrete choice experiment

Attributes	*Levels*	*1*	*2*	*3*	*4*
Packaging type	2	Standard	Local		
Price	4	-20% Market price	-10% Market price	+10% Market price	20% Market price
Brand	3	Focal brand	Secondary brand	Any other brand	

The levels and attributes were combined according to an L^MA design (Street, Burgess, Viney, & Louviere, 2008). Each participant was required to answer one version of a DCE with 16 choice sets, repeated for two product categories. Before starting the DCE, respondents were instructed to imagine a situation in which they were shopping, and the given options were the only products that were available on a supermarket shelf. In each choice set, the participants had to choose one out of four given options. Within each choice set, the position of brand alternatives was randomised. Consumers' choices were analysed using a multinomial logit model (Train, 2009). To estimate the relative importance of each attribute for this model, each attribute's partial R-square (the partial log-likelihood associated with each attribute across all of its levels) was calculated (Louviere, Islam, Wasi, Street, & Burgess, 2008). The part-worth utilities were estimated based on dummy variable coding and added up to zero. These were evaluated by means of z-scores and Wald statistics. Mathematically, part-worth is a partial utility generated by any given level of an attribute that can be added up to obtain the evaluation of a stimuli, such as a packaging attribute (Keller & Staelin, 1987). In DCE, it represents the extent to which the choice made by consumers is driven by a particular attribute.

Results of Attitudinal Measures

First, regarding the attitudinal part of the online questionnaire, the results for Chinese consumers were consistently similar to the ones in Chapter 6 for Pakistani consumers. That is, standard packaging (English) is preferred over localised (Chinese) packaging for hedonic products, regardless of the brand familiarity. For utilitarian products, similarly, there was no significant difference in the preference for a particular packaging type. These findings are presented in Table 7.2.

To determine how standard packaging and localised packaging might influence brand likeability, and whether their influence statistically differs from each other, we followed the procedure Cumming's (2009) has proposed. The results revealed a consistently significant influence of standard packaging likeability on brand likeability, but the results were mixed for localised packaging (see Table 7.3). For standard packaging, these findings strengthen the argument regarding the important role that packaging plays in enhancing product attractiveness and competitiveness.

Table 7.2 Paired statistics for mean ratings of standard and localised packaging likeability

Brands (n = 383)	*Mean (standard packaging)*	*Mean (localised packaging)*	*Mean difference*	*S.D*	*t-Value*	*p-Value*
Hedonic (food)						
M&M'S chocolate	5.43	5.28	0.14	0.98	2.01	0.04
Wonka chocolate	5.30	4.95	0.34	1.18	4.01	<0.01
Hedonic (non-food)						
L'Oréal nail colour	5.54	5.21	0.32	1.55	2.97	<0.01
Sally Hansen nail colour	5.16	4.94	0.21	1.17	2.54	0.01
Utilitarian (food)						
Maggi tomato sauce	5.32	5.26	0.05	0.89	0.90	0.36
McCormick tomato sauce	5.48	5.49	-0.01	0.87	-0.25	0.80
Utilitarian (non-food)						
Pantene shampoo	5.37	5.39	-0.02	0.92	-0.38	0.70
Nature's shampoo	4.94	4.92	0.02	0.87	0.41	0.68

Table 7.3 Effects of packaging likeability on brand likeability

	R	*R-square*	*S.E*	*Brand likeability (mean)*	*S.D*
M&M's	0.58[a]	0.33	0.82	4.76	1.07
	S.E	**Beta**	**Lower**	**Upper**	**Sig.**
Standard pack	0.08	0.51	0.28	0.71	<0.01
Localised pack	0.08	0.10	-0.03	0.40	0.23
Wonka	0.76[a]	0.57	0.65	4.32	1.16
	S.E	**Beta**	**Lower**	**Upper**	**Sig.**
Standard pack	0.10	0.30	0.12	0.51	<0.01
Localised pack	0.12	0.53	0.28	0.74	<0.01
L'Oréal	0.59[a]	0.35	0.81	4.82	0.98
	S.E	**Beta**	**Lower**	**Upper**	**Sig.**
Standard pack	0.06	0.53	0.33	0.72	<0.01
Localised pack	0.06	0.15	0.02	0.31	0.01
Sally Hansen	0.65[a]	0.42	0.76	4.18	1.08
	S.E	**Beta**	**Lower**	**Upper**	**Sig.**
Standard pack	0.05	0.43	0.23	0.62	<0.01
Localised pack	0.10	0.29	0.08	0.51	<0.01
	R	*R-square*	*S.E*	*Brand likeability (mean)*	*S.D*
Maggi	0.58[a]	0.34	0.82	4.40	1.01
	S.E	**Beta**	**Lower**	**Upper**	**Sig.**
Standard pack	0.09	0.44	0.10	0.67	<0.01
Localised pack	0.09	0.17	-0.05	0.54	0.06
McCormick	0.49[a]	0.24	0.87	4.63	0.93

Table 7.3 Cont.

	R	*R-square*	*S.E*	*Brand likeability (mean)*	*S.D*
	S.E	**Beta**	**Lower**	**Upper**	**Sig.**
Standard pack	0.09	0.21	0.01	0.40	0.01
Localised pack	0.08	0.32	0.15	0.55	<0.01
Pantene	0.66[a]	0.44	0.75	4.81	1.08
	S.E	**Beta**	**Lower**	**Upper**	**Sig.**
Standard pack	0.08	0.40	0.14	0.60	<0.01
Localised pack	0.08	0.31	0.10	0.60	<0.01
Natures	0.68[a]	0.47	0.73	4.11	1.25
	S.E	**Beta**	**Lower**	**Upper**	**Sig.**
Standard pack	0.12	0.42	0.20	0.64	<0.01
Localised pack	0.16	0.29	-0.01	0.63	0.08

Note

a Dependent variable: Brand Likeability.

Results of Discrete Choice Experiment

A DCE determined the most preferred attribute between price, brand and packaging as regards choosing a product. Using four product categories (chocolate, nail colour, shampoo and tomato sauce) at four price points and three types of brand familiarity (strong brand, weak brand or any other brand) for two different packaging types (standardised versus localised), we analysed the most preferred attribute in terms of choosing a product. Figure 7.1 shows an example of a discrete choice set of the experiment.

The results in Table 7.4 show that packaging (72%) is the most important factor for consumers' choice, followed by price (22%) and brand (5%), respectively, for the four product categories (chocolate, nail colour, shampoo and tomato sauce). The table also indicates that the part-worth for standard packaging is greater than that for localised packaging; therefore, consumers generally preferred the standardised pack to the localised pack.

Tables 7.5 and 7.6 present the results for the individual categories of hedonic products and utilitarian products, respectively. Across all categories, packaging type was the most important attribute (ranging from 50% for nail colour to 90% for tomato sauce) followed by price (ranging from 8% for tomato sauce to 34% for chocolate) and then brand (ranging from 0.6% for tomato sauce and 26.30% for nail colour). Packaging type was the most important attribute for hedonic products (i.e. chocolate and nail colour), and participants also preferred the standard packaging to the localised packaging, other than in the case of shampoo, for which they showed a clear preference for localised packaging.

Discussion

Overall, the findings extend the generalisability of the results of the study in Chapter 6 with Pakistani consumers. Here, it is similarly found with Chinese

Figure 7.1 Screenshot of discrete choice set.

Table 7.4 Results of discrete choice experiment

Attributes	*Part-worth*	*Wald*	*Wald %*	*p-Value*	*Mean*
Packaging type					
Standard	***0.62***	1325.11	**72.35**	<0.001	0.62
Localised	0.6				0.6
Any other brand	1.22				1.22
Brand					
Strong brand	***0.2***	97.87	5.34	<0.001	0.2
Weak brand	0.06				0.06
Any other brand	0.14				0.14
Price points					
p20	0.36	408.44	22.3	<0.001	0.36
p10	0.13				0.13
n10	0.19				0.19
n20	***0.29***				0.29
Sum of Wald		1831.42			

Notes

Adj. R^2 = 0.15; LL = -14,399.50; BIC (LL) = 28,840.64, N = 383; df = 376. Bold value represents the most important attribute. Bold and italic values represent the preferred attributes among the given levels.

Table 7.5 Discrete choice results for hedonic products

Chocolate	*Part-worth*	*Wald*	*Wald %*	*p-Value*	*Mean*
Packaging type					
Standard	***0.63***	294.42	***64.36***	<0.001	0.63
Localised	0.44				0.44
As it is	1.08				1.08
Brand					
Strong brand	***0.09***	4.64	1	0.09	0.09
Weak brand	0.1				0.09
Any other brand	0				0
Price points					
+20% Market price	0.41	158.39	34.62	<0.001	0.41
+10% Market price	0.22				0.22
10% Market price	0.25				0.25
20% Market price	***0.38***				0.38
Sum of Wald		457.45			

Notes
Adj. R^2 = 0.12; LL = 3630.95; BIC (LL) = 7298.49, N = 186; df = 179. Bold value represents the most important attribute. Bold and italic values represent the preferred attributes amongst the given levels.

Nail colour	*Part-worth*	*Wald*	*Wald %*	*p-Value*	*Mean*
Packaging type					
Standard	***0.61***	273.01	**50.01**	<0.001	0.61
Localised	0.6				0.6
As it is	1.21				1.21
Brand					
Strong brand	***0.51***	143.62	26.3	<0.001	0.51
Weak brand	0.2				0.2
Any other brand	0.3				0.3
Price points					
+20% Market price	0.36	129.26	23.67	<0.001	0.36
+10% Market price	0.22				0.22
10% Market price	0.26				0.26
20% Market price	***0.32***				0.32
Sum of Wald		545.89			

consumers that hedonic products should not be localised, whereas the packaging type does not matter for utilitarian products. The only exception is the choice of a utilitarian non-food product (DCE results of shampoo). These studies present evidence that localising the packaging of foreign brands may not necessarily be effective for Western brands in Asian markets. In this regard, these findings contrast with those of studies that contend that appealing to consumers' social identity is an effective marketing strategy (Özsomer, 2012; Zhou, Poon, & Wang, 2015). Conversely, the findings in both Chapter 5 and 6 suggest that localising

Table 7.6 Discrete choice results for utilitarian products

Tomato sauce	*Part-worth*	*Wald*	*Wald %*	*p-Value*	*Mean*
Packaging type					
Standard	***0.82***	518.02	**90.66**	<0.001	0.82
Localised	0.81				0.81
as it is	1.64				1.64
Brand					
Strong brand	0.07	3.41	0.6	0.18	0.07
Weak brand	***0.22***				0.22
Any other brand	0.14				0.14
Price points					
+20% Market price	0.21	49.92	8.74	<0.001	0.21
+10% Market price	0.14				0.14
10% Market price	0.17				0.17
20% Market price	***0.18***				0.18
Sum of Wald		571.35			

Notes
Adj. R^2 = 0.19; LL = 3337.29; BIC (LL) = 6711.24, N = 188; df = 181. Bold value represents the most important attribute. Bold and italic values represent the preferred attributes amongst the given levels.

Shampoo	*Part-worth*	*Wald*	*Wald %*	*p-Value*	*Mean*
Packaging type					
Standard	0.54	340.06	**69.55**	<0.001	0.54
Localised	***0.67***				0.67
As it is	1.21				1.21
Brand					
Strong brand	***0.23***	28.02	5.73	<0.001	0.23
Weak brand	0.29				0.29
Any other brand	0.06				0.06
Price points					
+20% Market price	0.48	120.91	24.72	<0.001	0.48
+10% Market price	0.05				0.05
10% Market price	0.11				0.11
20% Market price	***0.32***				0.32
Sum of Wald		488.99			

Notes
Adj. R^2 = 0.14; LL = 3690.86; BIC (LL) = 7418.59, N = 194; df = 187. Bold value represents the most important attribute. Bold and italic values represent the preferred attributes amongst the given levels.

the packaging of well-established Western brands of hedonic products may actually backfire.

Thus far, we have only considered product-side factors in terms of the packaging types, packaging elements or brand strength. In the study presented in the next chapter, we consider a consumer-side variable, the consumption situation—that is, whether or not conspicuous consumption occurs. In addition, following from the earlier chapters, it seems that even if standard packaging is preferred, particularly for hedonic products, questions remain regarding whether the product can be priced at a premium because the standard packaging may lead to more favourable perceptions of quality and value (Khan & Lee, 2020a). Hence, another aim of the next chapter is to investigate the effects of price premiums for brands in standard and localised packaging.

Note

1 Excerpts in this chapter are reprinted with permission from Emerald Publishing Limited; originally published in *Journal of Consumer Marketing*, *34*(5); ©Emerald Publishing Limited 2017.

8 Effects of Consumption Situation and Price Premium in Packaging Localisation

Influence of Consumption Situation

The concept of conspicuous consumption, first introduced by Veblen (1899), refers to the extent consumers choose a certain product to communicate their identities in a social situation (Barrera & Ponce, 2021; Neave, Tzemou, & Fastoso, 2020). By contrast, inconspicuous situation is about consumption in a private setting (Shove & Warde, 2002). Under the inconspicuous situation, consumers are often less concerned with being appreciated by others (Sullivan & Gershuny, 2004). Consequently, products consumed in the conspicuous situation generally hold higher symbolic value than those consumed in an inconspicuous situation. Drawing on symbolic theory (Cooley, 1902, products symbolise certain meanings. Therefore, consumers would use certain products to communicate these meanings (Zhou & Wong, 2008) or to present a desired self-image (Barrera & Ponce, 2021; Roy Chaudhuri, Mazumdar, & Ghoshal, 2011) to others present in the consumption situation.

Meanwhile, when people consume or possess products in social situations, product distinctiveness becomes a key factor for creating the desired impression of their selves on others (Snyder, 1992). The discernment of distinctiveness is emphasised both from the psychological and the social viewpoints, mainly because distinct products hold unique appeal (Lynn, 1992; Lynn & Harris, 1997). Hence, it is plausible that consumers' preference for a distinctive packaging emerges when the product is to be consumed under a conspicuous situation. Apart from using conspicuously consumed products to gain social appreciation, consumers may also use these cues to boost their self-esteem (Sivanathan & Pettit, 2010). For example, consumers often consume foreign products or products with foreign names to create an impression of being Westernised or modern (Doğan & Özkara, 2013).

However, the research findings for the inconspicuous situation are less certain. For example, Rahman, Zhu, and Liu (2008) assert that when the consumption situation is inconspicuous, consumers are less concerned about the symbolic value of consumption. In contrast, Eastman and Eastman (2015) suggest that the effects of product distinctiveness may also be preferred under the inconspicuous situation. Consumer desirability for distinct products consumed in conspicuous

DOI: 10.4324/9781003000822-10

situations is attributed to the signalling characteristics of products to denote affordability to others; and under inconspicuous situations, the aim is to fulfil personal gratification (Nelissen & Meijers, 2011). Further, brands' practices have been shifting societies from conspicuous to inconspicuous consumption, and thus, distinctiveness is not always important under conspicuous situations alone (Eckhardt, Belk, & Wilson, 2015; Kulviwat, Bruner II, & Al-Shuridah, 2009). This is because consumers may also prefer product exclusiveness in order to satisfice their own selves, rather than gaining social appraisals (Han, Nunes, & Drèze, 2010). Thus, these studies imply that unique packaging may be important under conspicuous as well as inconspicuous situations. Moreover, although distinctiveness is more important for conspicuous consumption, it is also important under inconspicuous situations (Khan, Lee, & Lockshin, 2017).

However, when distinctive products are consumed under public as well as private settings, the effects of social influence are mitigated or reduced in inconspicuous situations in comparison to conspicuous situations (Gao, Winterich, & Zhang, 2016). For example, one study shows the moderating role of public and private consumption in the purchase of exclusive products when consumed in a public setting (Kulviwat et al., 2009). The authors find a greater effect of social influence under conspicuous consumption than under inconspicuous situations. Therefore, distinctiveness becomes even more important under conspicuous situations than under inconspicuous modes of consumption (Khan, Lee, et al., 2017).

Influence of Price Premium

Expensive brands communicate affordability in a social context (Griskevicius et al., 2007). Consumption under the conspicuous mode is observable, and hence, consumers often purchase costly products to become noticeable to others (Amaldoss & Jain, 2005; Neave et al., 2020). When foreignness is used as a brand attribute in overseas markets, it would have a strong effect on consumers' purchase intentions because of the perceptions of high quality associated with these brands (Steenkamp et al., 2003). Such brands may attract a price premium to signal better quality because this enhances the perceived foreignness of the brand (Winit, Gregory, Cleveland, & Verlegh, 2014). Scarce or exclusive products are often favoured under conspicuous situations (Gierl & Huettl, 2010). Western packaging can be considered scarce in Asian markets owing to limited or no availability. As a result, consumers may relate with these brands better and may be willing to pay a higher price for these.

Consumers under conspicuous situations are also more willing to spend money for conspicuous products to signal to others about their prestige (Winit et al., 2014) and that they can afford these products (Mason, 1980), and they desire to be noticed by others (Barrera & Ponce, 2021; Chaudhuri & Majumdar, 2006; Neave et al., 2020). This is primarily because their behaviour is subject to social influence (Barrera & Ponce, 2021; Lessig & Park, 1982). However, the inconspicuous consumption motive is non-distinctive and free from social influence; therefore, the effects of cues regarding the country of origin may differ for such consumption (Piron, 2000). Under this consumption situation, consumers become less concerned about the symbolic value of their consumption (Rahman et al., 2008); thus, they may not be willing to pay a premium for their inconspicuous use when the same product is available at a lower price. Through an in-store study conducted in a supermarket in China, this chapter will shed light on how consumers' choices of standard versus localised packaging vary under conspicuous versus inconspicuous consumption situations. The study further investigates the effects of price premium on these choices.

Method: In-Store Choice Experiment

To test the variation in packaging likeability and the choice of respondents for standard versus localised packaging under conspicuous and inconspicuous consumption situations, each respondent of the study was primed to one of two different situations. The group of respondents who were exposed to the inconspicuous consumption situation was asked to imagine the following situation before choosing a product from the supermarket shelf: 'Other than you, no one is at home. You go to the supermarket to buy some products to take home for your own consumption'. Similarly, the other group was exposed to conspicuous consumption situation, whereby they were primed for the following situation: 'A group of friends visit your home. You go to the supermarket to buy these products to share with them'.

After they were primed, the respondents were asked to go to a supermarket shelf (labelled as Shelf A) on which they could find two products (M&M's, chocolate, i.e. food product and L'Oréal, nail colour, i.e. non-food product) in standard and localised packaging. Respondents were shown both types of packaging of these brands at the market price. Shelf A had no price difference between the standard and the localised packaging; later, Shelf B was used to test prices differences. The respondents were asked to choose and record in the questionnaire their purchase of either of the packs.

After they completed the choice experiment, the respondents were asked to answer attitudinal questions. They were also asked to rate the extent to which they liked the brands as well as the standard and the localised packaging of the brand on a 1–7 scale, anchored on *strongly dislike* to *strongly like*. Following the attitudinal ratings, respondents were asked to go to another shelf (labelled as Shelf B) on which they could find both products (M&M's, chocolate; and L'Oréal, nail colour) in standard and localised packaging. On this shelf, the respondents were presented the standard packaging of both brands at a price 15% higher than the market price, whereas they were presented the localised packaging at the market price. The reason for this step was to understand consumers' possible sensitivity to price premiums under conspicuous and inconspicuous situations. The respondents were requested to choose and record in the questionnaire their purchase of either of the packaging.

Last, respondents were required to answer a question pertaining to the manipulation check for the priming condition. The purpose of this question was to check whether the respondents were actually primed effectively to the given situation. For the conspicuous situation, the question was, 'For the product packaging that you previously selected from Shelf 1 and Shelf 2, do you think they are suitable for sharing with your friends at home?'. Similarly, for the inconspicuous group, the question was, 'For the product packaging that you previously selected from Shelf A and Shelf B, do you think they are suitable for your own consumption at home?'. The respondents were required to state their agreement or disagreement with the question through a yes or no response.

The final sample contained 130 participants, with 62 males and 68 females. The respondents' age ranged as a categorical variable from 20 to 51 years and above. They were divided into groups for answering the two questionnaires, with each questionnaire having roughly 65 participants. The conspicuous sample comprises 50% males and 50% females, and the inconspicuous group had approximately 46% males and 54% females.

Results

First, analyses were performed to determine attitudinal ratings. The results for the primed groups under the conspicuous and the inconspicuous situations showed significantly higher likeability for the standard packaging than for the localised packaging (see Table 8.1).

To further test whether the standard packaging likeability was greater under the conspicuous situation than under the inconspicuous situation, a paired sample *t*-test was performed on the standard packaging likeability for the two groups. The results for the differences in the standard packaging likeability confirmed that respondents assigned significantly higher ratings for standard packaging under conspicuous situations than under inconspicuous situations (see Table 8.2).

To analyse the choice of standard and localised packaging under the conspicuous and the inconspicuous situations, a non-parametric binomial choice test was performed. The test provided the percentage of respondents who selected standard and localised packaging. As Table 8.3 shows, the choice of packaging under the conspicuous situation remained significant for standard packaging, as did the choice of packaging under the inconspicuous situation. The choice of standard packaging was greater under the conspicuous situation than under the inconspicuous situation. Under the inconspicuous situation, the choice of standard packaging remained significant (68% for M&M's and 67% for L'Oréal), although the proportion of choice was still less than that for the conspicuous situation for both products (78% for M&M's and 88% for L'Oréal).

The results also evaluated the effects of price premiums under the conspicuous and inconspicuous situations (see Table 8.4). Even when the price was increased by 15% for the standard packaging and the localised packaging was available at the market price, respondents were still more likely to choose the standard packaging under the conspicuous situation. Under the inconspicuous situation, when the price was increased by 15% for the standard packaging and the localised packaging was available at the market price, respondents switched from the former

Table 8.1 Paired *t*-test statistics for standard and localised packaging likeability under conspicuous and inconspicuous situation

Brands (n = 64)	*Situation*	*Mean standard packaging*	*Mean localised packaging*	*Mean difference*	*t*	*S.D*	*S.E*	*Sig.*
M&M's chocolate	Conspicuous	5.87	5.10	0.76	3.71	1.64	0.20	<0.01
L'Oréal nail colour	Conspicuous	5.73	4.42	1.31	5.27	1.99	0.24	<0.01
Brands (n = 66)	*Situation*	*Mean standard packaging*	*Mean localised packaging*	*Mean difference*	*t*	*S.D*	*S.E*	*Sig.*
M&M's chocolate	Inconspicuous	5.25	4.81	0.43	1.93	1.84	0.22	0.05
L'Oréal nail colour	Inconspicuous	5.13	4.37	0.75	3.06	2.00	0.24	<0.01

Table 8.2 Paired *t*-test statistics for standard packaging likeability for conspicuous versus inconspicuous situation

Brands	*Mean conspicuous*	*Mean inconspicuous*	*Mean difference*	*S.E*	*t*	*Sig.*
M&M's chocolate	5.87	5.25	0.61	0.25	2.46	0.01
L'Oréal nail colour	5.73	5.13	0.59	0.26	2.26	0.02

Table 8.3 Choice of standard and localised packaging

Conspicuous situation (N = 64)	*Packaging choice*	*N*	*Obs. prop.*	*Test prop.*	*Sig.*
M&M's chocolate	Standard	50	0.78	0.50	<0.01
	Localised	14	0.22		
		64			
L'Oréal nail colour	Standard	56	0.88	0.50	<0.01
	Localised	8	0.13		
		64			
Inconspicuous situation (N = 66)	*Packaging choice*	*N*	*Obs. prop.*	*Test prop.*	*Sig.*
M&M's chocolate	Standard	45	0.68	0.50	<0.01
	Localised	21	0.32		
		66			
L'Oréal nail colour	Standard	44	0.67	0.50	<0.01
	Localised	22	0.33		
		66			

to the latter for the food product (M&M's), but not for the non-food cosmetic product (L'Oréal). That is, under the inconspicuous situation, respondents were still willing to pay a premium for the non-food cosmetic product, but not for the food product. Hence, they were sensitive to price premiums for the food product alone.

We further tested whether the influence of the standard packaging and the localised packaging on brand likeability differed statistically from each other using a procedure we adopted from Cumming (2009). First, for each product category, linear regression was performed with brand likeability as the dependent variable, and the ratings for the two packaging types as independent variables. The confidence intervals of the two regression coefficients were estimated via the bias corrected and accelerated (BCa) method using bootstrap. It is known that if two confidence intervals overlap by less than 50%, then the two regression coefficients are significantly different ($p < 0.05$). Table 8.5 shows that the regression coefficients of the standard packaging (β-M&M's = 0.34; β-L'Oréal = 0.41) were insignificantly different from the localised packaging (β-M&M's = 0.13;

Table 8.4 Choice of standard and localised packaging at different pricing levels

Conspicuous situation (N = 64)	*Packaging choice*	*N*	*Obs. prop.*	*Test prop.*	*Sig.*
M&M's chocolate at +15% market price	Standard	41	0.64	0.50	0.03
	Localised	23	0.36		
		64			
L'Oréal nail colour at +15% market price	Standard	41	0.64	0.50	0.03
	Localised	23	0.36		
		64			
Inconspicuous situation (N = 66)	*Packaging choice*	*N*	*Obs. prop.*	*Test prop.*	*Sig.*
M&M's chocolate at +15% market price	Standard	28	0.42	0.50	0.26
	Localised	38	0.58		
		66			
L'Oréal nail colour at +15% market price	Standard	45	0.68	0.50	<0.01
	Localised	21	0.32		
		66			

Table 8.5 Regression on packaging likeability and brand likeability

	R	*R-square*	*S.E*	*Brand likeability mean*	*S.D*
M&M's	0.39	0.15	0.93	5.36	1.47
	S.E	**Beta**	**Lower**	**Upper**	**Sig.**
Standard pack	0.10	0.34	0.12	0.54	<0.01
Localised pack	0.08	0.13	-0.06	0.31	0.11
	R	*R-square*	*S.E*	*Brand likeability mean*	*S.D*
L'Oréal	0.49	0.24	0.88	4.83	1.62
	S.E	**Beta**	**Lower**	**Upper**	**Sig.**
Standard pack	0.07	0.41	0.27	0.55	<0.01
Localised pack	0.09	0.23	0.06	0.39	0.01

β-L'Oréal = 0.23). The greater positive influence was found for the standard packaging on brand likeability for both brands.

Discussion

This chapter provides useful insights into packaging standardisation and localisation under conspicuous and inconspicuous modes of consumption. The results supported that standard packaging is effective for products under the conspicuous and the inconspicuous situations. Greater likeability and product choice under standard packaging is found under conspicuous situations, compared with inconspicuous situations. This finding is consistent with those of the studies that show modernity and product distinctiveness become more important when it is subject to social evaluation (Doğan & Özkara, 2013; Piron, 2000).

The findings also revealed that consumers were willing to pay more for the standard packaging when they purchase it for social consumption. In inconspicuous situations, they prefer the standard packaging, until they can find it at the same price as the localised packaging. When a price premium is charged on standard packaging, consumers switch to localised packaging in the case of a food product. However, they still prefer standard packaging for non-food cosmetic product. This may be because nail polish is categorised as a fashion product that is less sensitive to price premiums.

Consistent with the studies in the previous two chapters, this chapter's findings imply that marketers should not unconditionally or indiscriminately adapt product packaging for Asian markets. Brands in standard packaging can also attract price premiums. By adopting a standardisation practice in packaging for Asian markets, managers can gain greater market acceptability and benefit from premiums. Notably, this study only considers hedonic products and strong brands as stimulus. It adopts this approach because preliminary studies show that standard packaging is effective only for hedonic brands and that, regardless of whether the brand is strong or weak, consumers prefer standard packaging.

9 Language as a Graphic Element in Packaging Localisation[1]

Dual Role of Language as a Packaging Element

A packaging element may overlap more than one dimension. Language is a specific example of such a packaging element. When a product label is localised to the vehicular language of a country, it essentially serves as an informational element to aid cognitive processing (Hornikx & Meurs, 2021; Mueller & Szolnoki, 2010; Souiden et al., 2013). Fitzsimons et al. (2002) similarly contend that verbal cues induce efficient cognitive processing, which leads to effective consumer responses towards product evaluations. This contention resonates with research in marketing communication that shows that local language cues in advertisements are effective because consumers are able to better comprehend relevant information (Brumbaugh, 2002; Khan & Lee, 2020a; Koslow et al., 1994; Torres & Luna-Nevarez, 2012).

By contrast, research has shown that the mere display of a language, even when the language is not understood (e.g. see Gerritsen & Nickerson, 2010), can sway consumer behaviour (Khan & Lee, 2020a; Salciuviene et al., 2010). For example, brand names written in French enhance hedonistic characteristics (Leclerc et al., 1994), and brand names in English infer better quality to Chinese consumers (Li et al., 2012). Similarly, the use of the local language signifies national and cultural values (Zhou & Belk, 2004). Ostensibly, using a vehicular language leads to emotional attachment by consumers who self-identify with their own language (Krishna & Ahluwalia, 2008; Wagner & Charinsarn, 2021). Language cues heighten the salience of local consumers' self-identity, allowing them to associate with the brands, and help develop favourable responses towards the brands (Argo et al., 2006; Argo & White, 2012; Bishop et al., 2015). In these instances, the role of language is not to facilitate cognitive processing but to act as a symbolic cue to bias consumer perceptions, much like the function of graphic elements. Further examples of the symbolic, rather than the cognitive, role of language are listed in Table 9.1. In other words, language as a packaging cue can play either an informational or a graphic role.

DOI: 10.4324/9781003000822-11

Table 9.1 Studies on symbolism of foreign language

Source	*Language*	*Findings*
De Run (2005)	Different ethnic languages (Chinese, Malay and Iban) in advertisements	Malaysian consumers responded unfavourably to advertisements in a language incongruent with their ethnic background.
Hornikx, van Meurs, and Starren (2007)	French as foreign language in Dutch advertising	Foreign language (French) is associated with elegance and attractiveness.
Krishna and Ahluwalia (2008)	English versus Hindi language	English is found to be an effective strategy and implies modernity, whereas the native language provides sense of belongingness.
Micu and Coulter (2010)	English versus Romanian language	Regardless of the level of self-referencing, English language enhances modernity of the brand.
Holmquist and Cudmore (2013)	English language	The use of English language is considered novel and exotic, found to be appealing to consumers, regardless of language proficiency.
Nickerson and Camiciottoli (2013)	Arabic versus English language	English language as a threat to national language and identity among Arabic consumers.

Language Associations

As a form of human communication, language serves as a type of schematic lens for people to perceive the world around them (Hunt & Agnoli, 1991; Le Page, 1997; Santello, 2013). Indeed, as early as 1949, Sapir had surmised that language is a deliberate product of social habits that shapes how human beings think and perceive reality. This notion later became known as the Sapir—Whorf hypothesis (Hoijer, 1954; Hunt & Agnoli, 1991; Whorf, 1956), which similarly postulates that people's thinking and perceptions of the world only develop to the extent that the language that they speak allows it.

Consistent with the Sapir—Whorf hypothesis, sociolinguistic research suggests that language serves as a basis for symbolic construction, and people choose a particular language to convey their symbolic identity and to shape the perceptions they want to create about their message to the recipients (Alcántara-Pilar, Del Barrio-García, Porcu, & Crespo-Almendros, 2017; Feng & Adamson, 2018). For example, Blommaert (2010) contends that language evokes certain ethnocultural symbolic value due to specific attributes (e.g. being collectivistic or modern) attached with the people who speak the language. Consequently, a switch in the sociocultural framework arises when people use a language other than their vernacular language. Thus, a native Urdu speaker in Pakistan would adopt or convey a different cultural mindset when communicating in English as a second language, rather than in the native language.

Based on the sociolinguistic perspective of associations between a language and social attributes, research has linked English, particularly as a foreign language in non-Western countries, with modernity (Khan & Lee, 2020b; Lin & Wang, 2016). Lee (2006) asserts that this association originates from the polarised tussle between global discourses and local practices, such that the use of English has less to do with conveying the message than with portraying an image of being modern.

By contrast, among bilingual populations, the local language is inextricably linked to culture and serves as a common bond for self-identity (El Banna, Papadopoulos, Murphy, Rod, & Rojas-Méndez, 2018; Luna & Peracchio, 2005). Consequently, the local language is contrasted with the foreignness of English to symbolise unity and the sense of belonging (Koslow et al., 1994; Krishna & Ahluwalia, 2008; Lin & Wang, 2016). According to Micu and Coulter (2010), the link between a message in the local language and one's self-identity arises because the local language is likely to evoke self-referencing by relating the message to one's background. The association of a local language with the sense of belonging is analogous to research that affirms that language is an inherent component of one's ethnicity (Hornikx & Meurs, 2021; Laroche, Pons, & Richard, 2009; Luna & Peracchio, 2005). Supported by these findings, we envisage that in a non-Western bilingual country, English would be associated with concepts related to modernity, whereas the local language would be associated with concepts related to the sense of belonging.

The Markedness Model

The premise of Myers-Scotton's (1998, 2000, 2006) Markedness Model is that the choice of a language within a particular communication context brings with it the social characteristics or meanings associated with the country where the language is spoken. Consequently, the communicator would choose a certain language to portray a desired image or create a particular relationship with the recipient. Originally developed to explain the social motivations on why people switch codes, the model posits that a set of social codes prevails in a society to determine whether a linguistic choice is 'marked' (i.e. the language is unexpected) or 'unmarked' (i.e. the language is expected) in a communication context. The use of an unmarked language usually does not provoke salience, whereas a marked language may trigger strong salience and reactions due to its non-establishment within the context in which the language is used. Myers-Scotton (2006) also remarks that making a marked choice is a negotiation of who the speaker is in relationship to the other participants, and hence, it is a posturing of the solidarity or the power dimension, or both.

Within a marketing domain, studies have applied the Markedness Model to explain the responses of bilingual consumers to marketing messages (Alcántara-Pilar, del Barrio-García, Crespo-Almendros, & Porcu, 2017; Luna & Peracchio, 2005). For example, Krishna and Ahluwalia (2008) show that among bilingual consumers in India, the evaluations of advertisement slogans in English or Hindi (or both) differed depending on whether the advertisement was from a foreign or local company. The authors reasoned that the differences were due to the markedness of the language choice accentuating the attribute associated with the language (Hindi was the unexpected or marked language for the foreign firm, whereas it was English for the local firm). Later, Lin and Wang (2016) replicated Krishna and Ahluwalia's (2008) study, albeit in Taiwan with bilingual Chinese consumers, and found similar results for non-luxury products.

With product packaging, it is a common practice for foreign and local products to both have packaging in either English or local languages. Drawing on the Markedness Model, we argue that there will be an interaction effect between the country of origin of a product (i.e. English or local) and the language on the packaging (i.e. English or local); such that for a Western brand, the unmarked (i.e. expected) packaging language is English, whereas Urdu is the marked (i.e. unexpected) language. By contrast, for a domestic brand, Urdu is unmarked whereas English is marked.

Furthermore, in Markedness Model terms, a marked language makes the language stand out from its immediate context, thereby leading to the activation of a language schema to make the social meaning associated with the language more pronounced (Alcántara-Pilar, Del Barrio-García, Porcu, et al., 2017; Krishna & Ahluwalia, 2008). This phenomenon of a more pronounced association is likened to the concept of perceptual salience (Luna & Peracchio, 2005; Taylor & Fiske, 1975), the bias or tendency to use noticeable information to make judgements about an object. Hence, following on from the previous section regarding language associations, we contend that the association of English with modernity, or of Urdu with the sense of belonging, will be stronger if the language choice is marked (i.e. unexpected) in a given situation.

We further contend that language associations mediate the influence of language on packaging evaluation. This postulation is consistent with language schema theory (Luna & Peracchio, 2005; McVee, Dunsmore, & Gavelek, 2005), which suggests that the presentation of a language may activate a schema associated with the language. The evoked schema depends on the people who speak the language, the context in which it is used and recipients' perceptions and interpretations of the language (Myers-Scotton, 2006). Subsequently, the activated schema results in positive or negative assessment of the evaluated object. Krishna and Ahluwalia (2008; also see Lin & Wang, 2016) similarly argue that especially when the language in an advertising slogan is marked, people will first focus their attention on language-based associations and then use the associations to evaluate the slogan, which, in turn, influences their evaluation of the advertised object (Homer, 2006; Lau & Lee, 2018). Thus, we contend that for the localised packaging, the sense of belonging may mediate the influence of the packaging language on packaging likeability, and the effects are stronger for the Western brand than for the local brand. Similarly, for the English packaging, modernity mediates the influence of the packaging language on packaging likeability, and the effects are stronger for the local brand than for the Western brand.

Method

We chose Pakistan as the contextual market for this research. As an ex-British colony (prior to independence with India), the use of English is prevalent along with the local language, Urdu. Pakistani consumers are consumption oriented and spend significantly on the consumption of Western brands (Khan & Bamber, 2007). This is evident in supermarkets in major Pakistani metropolises, where supermarket shelves are well stocked with Western imported brands (Euromonitor, 2018). Less restrictive labelling regulations in Pakistan also facilitate the availability of different language versions of an identical product in the market (e.g. see Pakistan Country Guide, 2018).

Pretest 1—Brand Selection and Language Expectancy

The first task was to select a suitable product, which is gender-neutral (i.e. not likely to be primarily bought by one gender), with known foreign and local brands that are commonly available in major supermarkets and is available in English and localised packaging. Following a co-author's visits to major supermarkets in the Pakistani city of Lahore, we selected two brands of potato chips (*Pringles* as a foreign brand and *Chizzpa* as a local brand). Although there were possibly other options that would similarly meet these criteria (e.g. soft drinks or chocolate bars), the cost of running the main survey within an actual supermarket limited the number of products for this study to just one. In addition, the facing of potato chips' packaging is large relative to that of products such as soft drinks and chocolate bars, thereby ensuring that the language on the packaging was prominently noticeable by survey participants.

Data for the pretest were collected via a snowball sampling process. An email invitation (with an embedded link to an online English survey) was first sent to an academic in a top-tier Pakistani university. The email asked the recipient to complete the survey, before forwarding the email to ethnic Pakistani contacts, who are effectively bilingual in English and Urdu and may be interested in participating in the survey. Four days after the initial email, 49 responses were received. Since the number of responses remained unchanged for the next two days, we closed the online survey.

In the survey, the participants were shown four pictures of packaging, one at a time: Pringles as the Western brand in English packaging, Pringles in localised packaging, Chizzpa as the local brand in English packaging and Chizzpa in localised packaging. For each picture, the participants answered whether they believed the brand is of foreign or local origin. Then they were asked to examine the language on the packaging and answer (on a 7-point scale anchored on *highly unexpected* and *highly expected*) whether the language used on a particular packaging was expected or unexpected. Last, the participants reported their ability to read English (scale 1 = b*ad*; 2 = b*asic*; 3 = g*ood*; 4 = e*xcellent*).

The results showed that for Pringles, only one participant wrongly identified Pringles as a local brand, whereas seven wrongly identified Chizzpa to be a Western brand. Excluding these participants, the average scores for language expectancy were as followed: Pringles in English packaging (5.854); Pringles in localised packaging (2.286); Chizzpa in English packaging (2.289); Chizzpa in localised packaging (4.805). These results indicate that for the Western brand, English was more expected relative to the local language. Conversely, for the local brand, English was more unexpected than the local language. The participants were effectively bilingual, given the average English reading score of 3.384.

Pretest 2—Determining Language Associations

The purpose of the second pretest was to determine the characteristics that were associated with English and the local language, Urdu. Drawing on several studies that attempted to match a language with particular characteristics (Krishna & Ahluwalia, 2008; Lin & Wang, 2016), and following discussions among the authors to concur on their suitability, a list of ten attributes across a broad area was first developed (see Table 9.2).

Over three days, a male postgraduate Pakistani student approached potential participants (ethnic Pakistanis only) outside a supermarket in a popular shopping precinct in Lahore. The screening question was whether they would like and were capable of answering a paper-based questionnaire in English. Those who agreed were then asked to rate (on a 5-point scale anchored on *not associated at all* and *strongly associated*) the extent to which the language English and then Urdu were associated with the images or feelings represented by the ten attributes. The questionnaire did not contain pictures of any products. The participants also reported their gender, age group and English reading ability (scale 1 = b*ad*; 2 = b*asic*; 3 = *good*; 4 = *excellent*). The average English reading score was 3.23. The final sample size was 100.

We first conducted analysis of variance tests to eliminate attributes that might be biased by demographics (gender, age group and English reading ability); only 'Cosmopolitan' differed significantly in terms of English reading ability

Table 9.2 Mean scores of language associations

	Urdu#	*English*#	*Paired t-test*
Globalness	2.54	*4.13*	$t = -12.824$; $df = 98$; $p < 0.001$
Cosmopolitan	2.3	3.64	$t = -10.461$; $df = 98$; $p < 0.001$
Closeness	*3.94*	2.34	$t = 15.179$; $df = 98$; $p < 0.001$
Belonging	*4.34*	**1.81**	$t = 22.171$; $df = 98$; $p < 0.001$
Distant	2.82	2.95	$t = -0.918$; $df = 98$; $p = 0.361$
Caring	2.88	2.77	$t = 0.883$; $df = 98$; $p = 0.38$
Quality	2.03	*4.48*	$t = 15.286$; $df = 98$; $p < 0.001$
Credibility	2.28	3.75	$t = -10.586$; $df = 98$; $p < 0.001$
Professional	2.25	*3.92*	$t = -13.132$; $df = 98$; $p < 0.001$
Family	*3.85*	2.26	$t = 13.267$; $df = 98$; $p < 0.001$

Note
Italic/bold indicates the three highest scores.

($p < 0.001$) and was omitted in the main survey. 'Distant' and 'Caring' were also deleted for *t*-tests showed that their scores did not differ significantly between the two languages ($p = 0.38$ and 0.32, respectively). Of the seven remaining characteristics, we chose the three highest mean scores for each language. Thus, English was associated with 'Quality', 'Globalness' and 'Professional'; collectively, we termed these characteristics as *modernity*. Similarly, for Urdu, the collective term *sense of belonging* represented 'Family', 'Belonging' and 'Closeness'.

Main Survey Questionnaires

We adopted the between-subjects survey design for the product type (Pringles and Chizzpa) and the within-subjects survey design for the packaging type (English and local). All questionnaires were in English and were identical except for the product pictures. The first part of the questionnaire showed the picture of the English packaging, followed by questions regarding packaging likeability (rated on a 7-point scale anchored on *strongly dislike* and *strongly like*) and brand likeability (rated on a 7-point scale anchored on *strongly dislike* and *strongly like*). Next, the participants were asked to focus on the language on the packaging and rate (on a 5-point scale anchored on *not associated at all* and *strongly associated*) the extent to which the packaging language was associated with the images or feelings represented by the three attributes each for modernity and the sense of belonging. Last, they answered (on a 7-point scale anchored on *highly unexpected* and *highly expected*) whether the language used on the particular packaging was expected or unexpected. The second part of the questionnaire was identical to the first part except that the picture showed the localised packaging. To minimise potential order bias, the two parts were counterbalanced.

Last, participants reported whether they believed the brand was foreign or local, their familiarity with the brand (on a 7-point scale anchored on very unfamiliar and very familiar) and their gender, age group and English reading ability (rated on a 4-point scale with 1 = bad; 2 = basic; 3 = good; 4 = excellent).

Survey Participants

The main survey was conducted in a local supermarket in Lahore, where a shelf displayed Pringles and Chizzpa potato chips in both English and localised packaging. A trained field worker stood at the end of the shelf and randomly approached passing shoppers to participate in the survey. The participants were told that the questionnaire contained English and Urdu words, and that the survey was an academic study to understand consumers' perceptions of products. Those who agreed to participate were alternatingly assigned to the Pringles questionnaire or the Chizzpa questionnaire. In all, 209 participants completed the survey, with roughly equal numbers for both questionnaires. All participants were ethnic Pakistanis, and the English reading ability of the Chizzpa (mean score = 3.17 out of 4) and the Pringles (mean score = 3.23 out of 4) groups indicated that they are effectively bilingual.

Results

Language Associations

Table 9.3 shows the mean scores of associations between language and attributes for the Pringles and Chizzpa questionnaires. The results further affirmed the results of Pretest 2 that English was associated with modernity more than with the sense of belonging. Similarly, for both brands, Urdu was associated with the sense of belonging more than with modernity.

The Markedness Model

On comparing the localised packaging of the two brands, we find that the sense-of-belonging association was significantly stronger with the Western brand (4.11) than with the local brand (3.68); (t-value = 4.583; df = 206; $p < 0.001$). Similarly, comparison of the English packaging of the two brands shows that the modernity association was significantly weaker with the Western brand (3.41) than with the local brand (3.68); (t-value = -3.337; df = 206; $p = 0.001$).

To test the mediating influence of the language characteristics, we followed Baron and Kenny's (1986) procedure via three equations. Because consumers may use brand familiarity as a heuristic cue that serves as a shortcut to their choice preference, which would hence bias the outcomes of this study (Creusen et al., 2018; Heinberg, Ozkaya, & Taube, 2016), we included brand familiarity as a control variable in the model. The mean score of brand familiarity for Pringles (mean = 5.19) and Chizzpa (mean = 4.94) indicated that the participants are generally familiar with the brands.

$$PackLike = \beta_{10} + \beta_{11}\, BrandOrigin + \beta_{12}\, BF + \varepsilon_1 \qquad (9.1)$$

$$LangAssoc = \beta_{20} + \beta_{21}\, BrandOrigin + \varepsilon_2 \qquad (9.2)$$

$$PackLike = \beta_{30} + \beta_{31}\, BrandOrigin + \beta_{32}\, LangAssoc + \beta_{33}\, BF + \varepsilon_3 \qquad (9.3)$$

where
PackLike = packaging likeability; BrandOrigin = Western or local brand; LangAssoc = language association (either the sense of belonging or modernity); BF = brand familiarity.

Part A in Table 9.4 shows the results of running the regression analyses regarding the mediation of the sense of belonging with the localised packaging. Part B similarly shows the results of running the regression analyses regarding the mediation of modernity with the English packaging. In both parts, brand familiarity was not significant in all equations and their values are excluded from the table for parsimony (all $p > 0.106$).

In Part A, there was a treatment effect on the mediator (LangAssoc = sense of belonging) because β_{21} was significant. Moreover, the treatment of BrandOrigin

Table 9.3 Strength of language associations

	Local brand (Chizzpa), n = 103			*Western brand (Pringle), n = 106*		
	Urdu	*English*	*Paired t-test*	*Urdu*	*English*	*Paired t-test*
Packaging						
Globalness	2.45	3.59	$t = -7.536$; $df = 102$; $p < 0.001$	2.01	3.07	$t = -7.85$; $df = 105$; $p < 0.001$
Quality	2.49	3.77	$t = -12.479$; $df = 102$; $p < 0.001$	2.11	3.66	$t = -9.708$; $df = 105$; $p < 0.001$
Professional	2.9	4.01	$t = -6.782$; $df = 102$; $p < 0.001$	2.15	3.72	$t = -10.899$; $df = 105$; $p < 0.001$
Modernity	**2.62**	**3.68**	$t = -13.337$; $df = 102$; $p < 0.001$	**2.09**	**3.41**	$t = -16.711$; $df = 105$; $p < 0.001$
Closeness	3.49	3.09	$t = 2.785$; $df = 102$; $p = 0.006$	3.86	2.51	$t = 9.914$; $df = 105$; $p < 0.001$
Belonging	4.19	2.15	$t = 13.367$; $df = 102$; $p < 0.001$	4.43	1.91	$t = 20.554$; $df = 05$; $p < 0.001$
Family	3.37	2.63	$t = 3.87$; $df = 102$; $p < 0.001$	4.04	2.25	$t = 11.448$; $df = 105$; $p < 0.001$
Sense of belonging	**3.68**	**2.61**	$t = 11.706$; $df = 102$; $p < 0.001$	**4.11**	**2.22**	$t = 20.087$; $df = 105$; $p < 0.001$

Note
Modernity and ***Sense-of-belonging*** are super-categories, each with three attributes.

Table 9.4 Results of linear regression analyses

	Equation; dependent variable	*BrandOrigin*	*BrandOrigin *CET*	*LangAssoc*	*CET* LangAssoc*
Part A	1. PackLike	β_{11} = **0.305***	–	–	–
	2. LangAsc	β_{21} = **0.641****	–	–	–
	3. PackLike	β_{31} = 0.162	–	β_{32} = **0.217****	–
Part B	1. PackLike	β_{11} = **-0.495***	–	–	–
	2. LangAsc	β_{21} = **-0.454****	–	–	–
	3. PackLike	β_{31} = 0.162	–	β_{32} = **0.217****	–
Part C	4. PackLike	β_{41} = **-0.222***	β_{43} = 0.022	–	–
	5. LangAsc	β_{51} = **0.67****	β_{53} = -0.207	–	–
	6. PackLike	β_{61} = *-0.098*	β_{63} = -0.045	β_{64} = **0.229****	β_{65} = **0.135***
Part D	4. PackLike	β_{41} = **-0.298***	β_{43} = 0.021	–	–
	5. LangAsc	β_{51} = **-0.426***	β_{53} = -0.151	–	–
	6. PackLike	β_{61} = **-0.23***	β_{63} = -0.029	β_{64} = **0.201****	β_{65} = **-0.127***

Notes
PackLike = packaging likeability; BrandOrigin = Western or local; LangAssoc = language association (either sense of belonging or modernity); BF = brand familiarity. * $p < 0.05$; ** $p < 0.001$.

on PackLike turned from significant in β_{11} to insignificant in β_{31}. Collectively, these results indicated that the sense of belonging had mediated the effects of brand origin on the likeability of localised packaging. Further, the positive coefficient of β_{21} suggests that the mediating effects were stronger for the Western brand.

In Part B, there was a treatment effect on the mediator (LangAssoc = modernity) because β_{21} was significant. In addition, the treatment effect of BrandOrigin on PackLike turned from significant in β_{11} to insignificant in β_{31}. Collectively, these results indicate that modernity had mediated the effects of brand origin on the likeability of localised packaging. Further, the negative coefficient of β_{21} suggests that the mediating effects were stronger for the local brand.

A final noteworthy finding is that, as expected and consistent with other studies that suggest that packaging likeability leads to brand likeability (Creusen et al., 2018; Underwood, 2003), packaging likeability was positively correlated with brand likeability, although the relationships were weaker for the localised packaging of both brands (significant at the 0.01 level only). The correlation values for the local brand were as follows: English packaging (r = 0.491; $p < 0.001$) and localised packaging (r = 0.172; p = 0.082). Similarly, the correlation values for the Western brand: English packaging (r = 0.449; $p < 0.001$) and localised packaging (r = 0.18; p = 0.062). Last, since the demographic characteristics (age, gender, English reading ability) did not correlate with the sense of belonging or modernity, it means that the effects of the language associations on packaging likeability may be broad-based.

Discussion

The purpose of this study is to shed light on the role of language, as a graphic packaging element, in influencing consumers' evaluation of product packaging. Drawing on sociolinguistic studies, particularly the Markedness Model, it investigates how the markedness (i.e. expectancy) of different languages on products from a different country of origin may be a factor in guiding the evaluation. Collectively, the finding that the localised packaging draws out a strong association with the sense of belonging suggests that brands that wish to appeal to the common ethnic bond among local consumers should use the local language on their packaging. Similarly, using English as the packaging language would allow brands to position themselves as modern and progressive.

The mediating effects of the language associations infer a successively linked process, in which the presentation of a language first gives rise to social characteristics associated with the language, which, in turn, influence the evaluation of an object. The effects are consistent with the assertion that a presented language first activates a schema, and the information retrieved from the schema is then used to interpret the context within which the language is presented (Melendez & Pritchard, 1985; Myers-Scotton, 2006). Particularly for branding research, this means that stereotypical characteristics associated with a language may be transferred to the product in question, and hence, brands should pick the appropriate language depending on the characteristics they wish to convey. More importantly, brands have to consider the markedness of the chosen language because this will accentuate the associated characteristics, if this is what the brands wish to achieve.

In all studies presented in the preceding chapters thus far, we have not considered any individual dispositional variables. In particular, we are interested in understanding the effects of the need for cognition (NFC; Cacioppo & Petty, 1982) and the need for affect (NFA; Maio & Esses, 2001) as individual difference variables. An important reason for considering these variables is because research has shown that they are associated with the effectiveness of affect- and cognition-based cues (Haddock, Maio, Arnold, & Huskinson, 2008). In packaging design, as we have discussed in Chapters 3 and 4, information elements are processed cognitively, whereas graphic elements are likely to involve affective processing. Given that individual consumers may possess different NFC traits or NFA traits, they are likely to evaluate packaging differently, and the next chapter serves to shed light on this issue.

Note

1 Excerpts in this chapter are reprinted with permission from Palgrave MacMillan, and were originally published in *Journal of Brand Management*, *27*(2), © Palgrave MacMillan 2020.

10 How Consumers' Psychological Dispositions of Need for Affect and Need for Cognition May Influence Packaging Localisation Effectiveness

Wayne Wang, Susan Freeman, Anne Sharp and Song Yang

As we have discussed in Chapters 3 and 4, packaging elements may be informational (i.e. cognitive stimuli) or graphic (i.e. affective stimuli). In this regard, it is likely that consumers' individual differences in the *need for cognition* (*NFC*; Cacioppo & Petty, 1982; Rathore, Jakhar, Kumar, & Kumar, 2022) and the *need for affect* (*NFA*; Diamantopoulos, Arslanagic-Kalajdzic, & Moschik, 2020; Maio & Esses, 2001) may interfere with their evaluations of product packaging. That is, NFC and NFA may colour the perceptions of cognition-based and affect-based cues in packaging (Haddock et al., 2008). Thus, in considering these individual dispositional effects, this chapter reports the findings of two studies, both conducted in China. Study 1 executes a discrete choice experiment (DCE) to test the differential influence of individual differences in NFC and NFA (Chang & Yen, 2013; Maio & Esses, 2001). Given that consumers often shop under time pressure, Study 2 mimics this condition and experimentally demonstrates how time pressure may alter Study 1's results.

Study 1: Individual Consumer Differences, Brand Familiarity and Localisation Strategies

Need for Cognition and Need for Affect

Cohen, Stotland, and Wolfe (1955) defined the *NFC* as 'a need to structure relevant situations in meaningful, integrated ways. It is a need to understand and make reasonable in the experiential world' (p. 291). NFC reflects the individual's motivation in processing information, and high-NFC individuals tend to engage in effortful activities and seek out cognitive information when faced with choice decisions (Cacioppo & Petty, 1982; King & Auschaitrakul, 2021; Tu, Kwon, & Gao, 2021). Cacioppo, Petty, and Feng Kao (1984) developed a 34-item NFC scale and linked it to individual motivation in processing information, namely effortful cognitive endeavours. Later studies

DOI: 10.4324/9781003000822-12

attempted to reduce the number of items in the scale. For example, Verplanken (1991) employed a 6-item version of the NFC scale to measure the persuasion effectiveness of risk information.

The NFC scale indicates that individuals with high scores on this scale tend to enjoy thinking about, and engaging in, effortful cognitive activities. This type of individuals also attempts to seek out cognitive messages and be thoughtful before making decisions. For instance, Haugtvedt, Petty, and Cacioppo (1992) reveal that high-NFC individuals prefer attitude-based evaluations of product-related attributes in advertisements. In contrast, individuals scoring relatively low on the scale prefer to avoid cognitive or complex activities. These individuals often have low motivation to process cognitive messages and are more likely to seek out tasks with simple rules. As discussed in Chapters 3 and 4, informational elements on packaging aim to provide product-related information and support consumers' cognitive responses. Therefore, Study 2 argues that individual differences in the NFC moderate the persuasion effectiveness of informational elements in packaging.

Apart from their NFC, people also have an innate desire to feel and pursue emotions (Haddock et al., 2008; Sojka & Giese, 1997). The *NFA* refers to the tendency, desire and motivation of an individual to approach or avoid situations and activities that are emotional or affective or include this type of information (Maio & Esses, 2001). It is a tendency to associate feelings and emotions, rather than objective assessments, with messages (Huskinson & Haddock, 2004). For example, some people tend to prefer activities with affective content (e.g. emotional movies) over activities without affective content (e.g. non-emotional movies; Haddock et al., 2008). Consequently, high-NFA individuals tend to seek affective objects or are easily influenced by an object's emotional attributes (i.e. graphic packaging elements).

The NFA scale was developed for understanding the affective intensity of information based on an individual's disposition. This scale includes 26 items for measuring the approach-or-avoid affective attitudes (Maio & Esses, 2001). It indicates that individuals with high scores are more willing to associate their feelings and emotions with the messages on the stimulus for shaping their judgements and choice behaviour. In contrast, it indicates that individuals with low scores tend to hide their feelings/emotions rather than expressing or connecting to the perceived object (Maio & Esses, 2001). Further, high-NFA individuals have a strong desire to seek out affective information on the object, unlike low-NFA individuals, who lack that motivation (Maio & Esses, 2001). Later, Appel, Gnambs, and Maio (2012) developed a shorter, more efficient 10-item version of this scale.

Need for Cognition and Need for Affect May Influence Packaging Evaluation

Taken together, the review in the previous section shows that individual differences in NFC and NFA can play essential roles in measuring the effectiveness of persuasion information. Packaging as a communication tool involves both cognitive (informational) and affective (graphic) elements. It thus stands to reason that NFC and NFA may moderate the persuasion effects of packaging, especially when packaging elements are modified for non-Western markets. In this sense, the differences between NFC and NFA resemble the central route versus peripheral route of persuasion in the Elaboration Likelihood Model (Petty & Cacioppo, 1986).

Furthermore, in NFA and NFC research, it is important to account for brand familiarity, which refers to consumer's knowledge about a brand that arises from their direct and indirect experience with the brand (Campbell & Keller, 2003). This is because brand familiarity has a dominant influence on how consumers use their stored brand knowledge as a heuristic cue for less effortful decision-making (Hoyer & Brown, 1990). When consumers are unfamiliar with a brand, they are likely to want to learn more about the brand in order to generate an initial brand impression (Campbell & Keller, 2003; Hilton & Darley, 1991). Such a situation occurs when a foreign brand is introduced into a new market, as is the case in market expansion to foreign countries. In contrast, Baker, Hutchinson, Moore, and Nedungadi (1986) propose that high brand familiarity may increase consumer affective responses to the brand. This means a product with higher brand familiarity requests less effortful evaluation by consumers in the information process.

On the basis of this review, it would be interesting to determine the interaction among NFC or NFA, brand familiarity and the preference for different packaging elements (i.e. informational versus graphic). As regards familiar brands, there may be no difference between high-NFC and low-NFC individuals in their preference for informational elements.

Method

This study was conducted in China because the unique Chinese culture and language and its strong consumption landscape mean that the country is a suitable representation of a non-Western market for this study (Batra, Ramaswamy, Alden, Steenkamp, & Ramachander, 2000). First, we conducted a pretest to identify suitable products and packaging-element stimuli for the main experiment. For the pretest, we first selected six consumer packaged goods (CPG) categories, each with one brand, as potential candidates for the main study. The categories (brands) were potato chips (Smith), soft drinks (7 Up), shampoo (Dove), chocolate (Snickers), cleaning detergent (Vanish) and adult milk powder (a2).

Fifty-two participants recruited via an online panel in China answered an online questionnaire. The first part of the questionnaire displayed the pictures of the six products, captured by digital photography, as they appeared in the Chinese supermarkets. We only showed the front side of the packaging because prior research has established that 90% of consumers only examine the front side of the packaging before making a choice (Urbany, Dickson, & Kalapurakal, 1996). The participants were asked to rate the extent to which they were familiar with each brand on a 7-point semantic differential scale anchored on *unfamiliar* (1) and *familiar* (7). The results of the pretest test indicated that Snickers chocolate (6.4) and a2 adult milk powder (2.08) were the most and the least familiar brands, respectively. For the packaging elements, the product description (PD) and the product name (PN) were identified as informational elements, whereas the brand name/tagline and cultural cues were viewed as graphic elements.

The participants also answered the NFC and NFA scales. Drawing on Verplanken and Pieters (1988), the 6-item NFC scale measured individual motivation to engage in effortful activities and seek out cognitive information. Similarly, the 5-item NFA scale measured individual tendency to associate feelings and emotions, rather than objective assessments, with messages (Appel et al., 2012). Two bilingual researchers used a standard back-translation procedure to translate all the scale items into Chinese (Craig & Douglas, 2005). All items used a 7-point scale anchored on *strongly disagree* and *strongly agree.*

The main survey was also conducted online via a panel provider. It comprised two parts. In the first part, respondents participated in a DCE to capture their choice preferences. Similar to the pretest, the two chosen products (Snickers chocolate and a2 milk powder) were digitally created on the online questionnaire. The DCE design had three attributes (PD, brand name/tagline and cultural cues). The first two attributes had two levels each, one as the attribute appeared in its standard (English) packaging and the other as it appeared in the localised packaging. The two levels of cultural cues were the presence or absence of a visual cue; the visual cue for Snickers chocolate was a Chinese model (see Green [1999] regarding the use of an ethnic model as a cultural symbol), whereas a bamboo plant was the cue for a2 milk powder, which symbolises longevity and nobility in the Chinese culture (see Yetts [1912] regarding the use of the bamboo plant as cultural symbol).

Hence, there were 16 possible combinations for the packaging of each brand. Following the design criteria developed by Huber and Zwerina (1996), we achieved orthogonality, level balance, minimum overlap and utility balance in the designing process, with the efficiency of the proposed choice sets at 82.4%. Each respondent completed ten discrete choice tasks (five sets for each product), all of which were binary choice tasks. Although no time limit was set for task completion, the time spent on each choice set was recorded in order to test the hypotheses for Study 3.

In the second part of the online survey, respondents reported their familiarity with the two brands (using the same 1–7 scale as in the pretest). They then answered the NFC and NFA scales. A total of 145 participants completed the main survey. Participant ages ranged from 18 to 65 years ($M = 33.7$), with 60% male and 40% female. We used a Likert scale to assess their English proficiency, ranging from 1 (*poor*) to 5 (*good*). The average score for respondents' English proficiency was 2.7.

Results

The analyses for the NFC and NFA factors indicated that both factors possessed adequate reliability; Cronbach's α was 0.90 for NFC and 0.84 for

NFA. Convergent validity was also adequate because all factor loadings were greater than 0.5 (Campbell & Fiske, 1959). The average variance extracted (AVE) estimates for the two factors were higher than the correlation between the factors (r = 0.378, AVE = 0.58), thereby confirming discriminant validity (Fornell & Larcker, 1981). We also performed Harman's one-factor test to check for multicollinearity problems. The variance of one factor (40.86%) was less than the cumulative variance of each factor in the exploratory factor analysis (58.60%). Thus, it was confirmed that there was no problem of multicollinearity (Podsakoff, MacKenzie, Lee, & Podsakoff, 2003). Last, discriminant validity was also supported by fitting the data to a measurement model with both constructs (CFI = 0.98; IFI = 0.98; RMSEA = 0.052; χ^2 [57.5/43] = 1.34, p = 0.067).

Brand familiarity scores were 5.68 for Snickers chocolate and 3.16 for a2 milk powder, similar to the pretest results, and supported our use of Snickers as the familiar brand and a2 as the unfamiliar brand. To test the effects of NFC and NFA on packaging evaluation, we first dichotomised NFA into high-NFA and low-NFA based on their factor scores, and similarly dichotomised NFC.

Since we were comparing the choice preference for a localised versus a Western version of a product, consumer ethnocentrism may have biased the choice decisions (Lee, Lee, & Li, 2017; Strizhakova & Coulter, 2019). We operationalised consumer ethnocentrism (Cronbach's α = 0.81) using a 6-item scale (Lee et al., 2017), and then tested whether the respondents' consumer ethnocentric dispositions differed between those who chose the localised packaging versus those who chose the standard packaging for each of the ten choice sets. The t-test results showed that consumer ethnocentrism was not significant in any of ten choice sets (all $p > 0.917$). Hence, we concluded that consumer ethnocentrism did not bias the choice decisions and excluded it from further analyses.

Next, we analysed the discrete choice data using the JMP by SAS software (version Pro 12.1) choice-based multinomial logit model. The levels of the attributes were coded as dummy variables, so that the utilities of each attribute would add up to zero. In particular, localised attributes were coded as (-1), whereas standard attributes were coded as (1). This model enabled us to measure the Wald chi-square statistic and parameter estimates. The Wald chi-square can reveal the importance of each attribute to the respondents, whereas more importantly, the parameter estimates can reflect the effects of the localised attributes. Next, we captured the utility of localised attributes. If the utility value shows a positive number, it means the participants prefer localised attributes over the similar attributes in the standard version. In other words, localised attributes have a greater impact on the choice decisions than the attributes in the standard version.

High-NFC Individuals' Choice Preference

As shown in Table 10.1, for the unfamiliar brand, the localised informational elements (PN, PD) on the packaging dominated the choice preference

Table 10.1 Parameter estimates and chi-square of discrete choice experiment

Subjects	*Brand familiarity*	*Attribute in localisation*	*Estimate*	*Chi-square*
High-NFC	Familiar brand	PN = Product name (L)	0.159***	5.183
		PD = Product description (L)	0.001	0.054
		BT = Brand name and tagline (L)	-0.252**	11.714
		CC = Cultural cues (L)	0.002	1.022
High-NFC	Unfamiliar brand	PN = Product name (L)	0.278*	16.453
		PD = Product description (L)	0.357*	27.198
		BT = Brand name and tagline (L)	-0.329*	30.728
		CC = Cultural cues (L)	-0.028	0.7191
High-NFA	Familiar brand	PN = Product name (L)	0.057	0.635
		PD = Product description (L)	0.006	0.011
		BT = Brand name and tagline (L)	-0.257*	17.558
		CC = Cultural cues (L)	-0.084	1.316
High-NFA	Unfamiliar brand	PN = Product name (L)	0.048	0.316
		PD = Product description (L)	0.002	0.003
		BT = Brand name and tagline (L)	-0.475*	55.777
		CC = Cultural cues (L)	-0.108	0.102

Notes
L = Localised. * $p < 0.001$; ** $p < 0.01$; *** $p < 0.05$.

of the high-NFC respondents (β-PN = 0.278, Wald-PN = 16.453, $p < 0.001$; β-PD = 0.357, Wald-PD = 27.198, $p < 0.001$). This result supports the postulation that when high-NFC individuals are choosing the packaging of an unfamiliar brand, localised informational elements have a greater impact on their choice decisions than standard informational elements. However, the results also showed that the standard brand name and tagline (BT) has a strong persuasion effect on the high-NFC participants' choice (β-BT = -0.329, Wald-BT = 30.728, $p < 0.001$), but the cultural cues in both versions have no significant choice impact.

A DCE can present the utility of each level and allows researchers to combine levels into various random combinations to make trade-offs (Cohen, 2009). Thus, for high-NFC consumers, the combination with the strongest persuasion effects for low-familiarity brands packaging is:

PN(L)* + PD(L)* + BT(S)* + CC(S) with the utility value = 0.278
(*$p < 0.001$; **$p < 0.01$; ***$p < 0.05$)
(L = Localised; S = Standard)

For the familiar brand, the persuasion effects of the two informational elements in localisation showed variance. In particular, high-NFC individuals still preferred the PN in localisation than the standard version, whereas PD had no significant impact. Meanwhile, within graphic elements, the brand name and tagline in standardisation showed the strongest choice impact among the four attributes, whereas localised cultural cues were slightly more preferred (but not significantly) than the non-localised cultural cues. Taken together, the results showed that for the familiar brand, high-NFA individuals preferred the localised PN and the standard brand name and tagline over the standard PN and localised brand name and tagline. Meanwhile, the standard brand name and taglines have the strongest impact on the choice decisions. Therefore, this result is inconsistent with H1b. Similarly, for high-NFC consumers, the combination with the strongest persuasion effects for high-familiarity brands packaging is:

PN(L)*** + PD(L) + BT(S)** + CC(L) with the utility value = -0.09
(*$p < 0.001$; **$p < 0.01$; ***$p < 0.05$)
(L = Localised; S = Standard)

Last, high-NFC respondents spent a longer time (on average, 11.8 seconds per choice set) on the choice sets for the unfamiliar brand than on those for the familiar brand (on average, 8.9 seconds per choice set).

High-NFA Individuals' Choice Preference

For the unfamiliar brand, unsurprisingly, the brand name and tagline of the standard version dominated high-NFA respondent choice preference. Meanwhile, high-NFA individuals preferred localised cultural cues slightly more than they did the same elements in the standard version; however, the effect is insignificant. The informational elements did not significantly affect the choice decisions either. This result partially supports H2a. For high-NFA consumers, the combination with the strongest persuasion effects for low-familiarity brand packaging is:

PN(L) + PD(L) + BT(S)* + CC(S) with utility value = -0.533
(*$p < 0.001$; **$p < 0.01$; ***$p < 0.05$)
(L = Localised; S = Standard)

For the familiar brand, this study presents the same results as for the unfamiliar brand in the case of high-NFA respondents. The results showed that the brand name and tagline in the standard version has a significant impact on the choice preference, but the remaining three attributes in both versions had insignificant effects. Thus, the result partially supports H2b. Therefore, for high-NFA consumers, the combination with the strongest persuasion effects for high-familiarity brands packaging is:

PN(L) + PD(L) + BT(S)* + CC(S) with utility value = -0.278
(*$p < 0.001$; **$p < 0.01$; ***$p < 0.05$)
(L = Localised; S = Standard)

Last, high-NFA respondents spent a slightly longer time (on average, 6.8 seconds per choice set) on the choice sets for the unfamiliar brand than on those for the familiar brand (on average, 6.3 seconds per choice set).

Discussion

Based on the DCE results, we find that the informational dimension in the localised packaging has a stronger influence than this dimension in the standard packaging when high-NFC consumers are choosing low-familiarity brands. That is, first, compared with the high-familiarity brands, consumers tend to learn more about the brand and the product in order to generate judgements (Campbell & Keller, 2003; Hilton & Darley, 1991). Second, compared with the consumers with other dispositions, high-NFC individuals' choice decisions and behaviour are more likely to be influenced by cognitive information (Cacioppo & Petty, 1982; Verplanken, Hazenberg, & Palenewen, 1992). This finding is consistent with that of Haugtvedt et al. (1992), who show that high-NFC individuals are more likely to shape choice behaviours based on processing product-related attributes, whereas this was not the case for those with other dispositions. Especially when the information was written in the local language, it provided a stronger impact on cognitive thinking than the standard informational dimension. Furthermore, the time taken to complete the online DCE revealed that high-NFC respondents spent more time on evaluating the packaging of the unfamiliar brand than the familiar brand. These results enhance the robustness of the previous findings.

However, the high-NFC participants also preferred the standard brand name and tagline on the package, even though they were unfamiliar with the brand. This phenomenon may be specific to the context of this study—Chinese consumers—and can be explained as follows. It is plausible that for Chinese consumers, foreign brand names and taglines may symbolise reliable quality, Westernisation and modernity, resulting in their preference for the standard graphic elements (Batra & Homer, 2004; Dawar & Parker, 1994). For these reasons, although the localised informational elements support the cognitive efforts of high-NFC consumers, the brand cues in the standard version may also contribute certain influences on choice making. This finding is also consistent with that of Huskinson and Haddock (2004), who argued that some people may employ both cognitive and affective information to an equal degree. In contrast, although the standard brand name and tagline significantly affected the high-NFC participants' choice decision, the importance weight of the attributes was not as strong as the weight of the localised PD on the package. That is, the PD in the local language still primarily influences high-NFC individual choice, because the element often contains the most product-related information among all the elements on the packaging.

For the familiar brand, the results showed that for the informational dimension, high-NFC consumers preferred the localised PN more than the PN in English. Meanwhile, for the graphic dimension, the standard brand names and taglines had a stronger impact on the high-NFC consumers' choice decisions than the similar elements in the localised version. The possible explanations for these findings are as follows: First, because of the high brand familiarity, the standard brand cues in the graphic dimension may reduce consumers' cognitive efforts on the informational dimension (Baker et al., 1986). In particular, the high-NFC respondents were no longer processing the localised PD in the informational dimension to generate their choice behaviour. Second, as Shiv and Fedorikhin (1999) proposed, 'people's evaluation could arise from stimulus-based as well as memory-based processes and could be both favourable toward the alternatives in the choice task' (p. 280). It implies that when high-NFC individuals are choosing familiar brands, although they tend to examine the localised PN to support their cognitive thinking, they may still consciously utilise their memory to make choices. This is because the memory-based process is more controlled and seeking out cognitive messages is the nature of high-NFC individuals (Cacioppo et al., 1984; Lynch, 1991; Romaniuk & Sharp, 2015).

For the high-NFA individuals, the results showed that only part of the elements in the graphic dimension significantly influenced their choice making. In particular, high-NFA respondents were more likely to choose the packaging with the standard brand name and tagline than the localised element, whereas the cultural cues in both versions insignificantly influence the choice decisions. The standard brand cues in the graphic dimension may have significantly influenced high-NFA consumers' choice decisions for the following reasons. First, high-NFA individuals are more likely to be influenced by affective messages (the graphic dimension in this case; Appel & Richter, 2010; Maio & Esses, 2001).

Second, for the unfamiliar brand, the standard brand cues may deliver symbolic meanings such as Westernisation and modernity, that are associated with high-NFA respondents' social and self-belonging (Batra & Homer, 2004; Brooks, 1994). Although the respondents have less brand knowledge and consumption experience of the brand, the symbolic meanings of these cues may also influence high-NFA individuals' choice behaviour. In contrast, when high-NFA respondents were familiar with the brand, they were more likely to retrieve standard brand cues from their memory, which affectively influence their choices (Keller, 2003; Romaniuk & Sharp, 2015).

Although the standard brand cues of familiar and unfamiliar brands significantly influenced high-NFA respondents' choice making, the standard brand cues on the packaging of the unfamiliar brand had a stronger impact than did these cues on the familiar brand. A reason may be the product type used in this study coupled with the context of China. Chinese consumers are more concerned about the quality of the milk powder than the chocolate bar because the entire family including children, potentially, consume the milk powder. Here, the standard graphic dimension can signal product quality (Espejel, Fandos, & Flavián, 2009). Regarding Western brands, they are often perceived as having a more reliable quality, which is the key attribute that non-Western consumers seek (Khan et al., 2015). Hence, high-NFA respondents were more likely to be influenced by the standard brand cues of the milk powder (the unfamiliar brand) than the similar elements of the chocolate bar (the familiar brand) in this study.

Surprisingly, the cultural cues in this study did not have any impact on the choice decisions of the respondents. In contrast, many studies have emphasised the symbolic value of cultural cues in various marketing domains. Especially in the non-Western markets, consumers' choice behaviours are often influenced by the symbolic meaning of cultural attributes (Bakar et al., 2013; He & Wang, 2017; Zhang, 2014). However, this study revealed that cultural cues on CPG packaging are less important. The major reason may be that the brand originates from the Western market, and hence, consumers were not interested in any localised culture cues attached to the packages of the products. Therefore, it is unnecessary to modify the cultural cues on packages for non-Western markets.

Overall, these findings provide several managerial implications. The results indicate that the effectiveness of packaging design for non-Western markets may be improved by understanding the differences among individual consumers. For a target group, such as high-NFC or high-NFA individuals, marketers of Western CPG brands should carefully modify their product packaging in non-Western markets. Interestingly, a McKinsey report suggested that marketers should focus on promoting the affective-based attributes of CPG since 24% of Chinese consumers are driven by their emotions to make choices (Atsmon & Magni, 2012). This report enhanced the importance of the need-for-affect concept revealed in this study.

The findings also contribute insightful knowledge to managers of both well-known brands (with high brand familiarity) and new brands (with low brand familiarity) in developing an effective packaging localisation strategy for non-Western markets. Specifically, combining the moderating effects of individual differences, this study showed the effectiveness of various combinations of packaging design in localised and non-localised versions for both well-established and new brands when sold in non-Western markets. Thus, it sheds new light to guide Western brands in developing effective packaging localisation strategies for non-Western markets. That is, although most CPG companies aim to increase brand awareness globally, this study delivers an alternative insight, namely, that before achieving this goal, they need to modify the packaging effectively to approach targeted groups.

Study 2: Choices under Limited Time

The purpose of Study 2 is to extend Study 1's findings by imposing a time limit during the DCEs to better represent the time constraints that shoppers often face when shopping for consumer goods.

Method

The experimental design and procedure of Study 2 were identical to that for Study 1. The only difference was that we constrained the time on the choice sets. Each respondent was given five seconds to complete a choice task. The respondent was automatically forwarded to the next choice set when the given time ran out. The actual time spent was captured. The same online panel used for Study 1 provided 125 participants for this study. After excluding respondents with missing data, the final sample comprised 104 respondents. Their age range was 18–65 years; 42% were males and 58% were females. About 21% identified themselves as having good English reading ability. None of these participants had participated in the Study 1 experiment.

Results

To ensure the resource availability was successfully manipulated, the actual time spent for each set from both experiments was compared. A significant difference was found in the perceived choice time spent for the unfamiliar brand ($F = 2.984$, $p < 0.001$) and the familiar brand ($F = 3.104$, $p < 0.001$) in the two experiments. The Cronbach's α for the 6-item NFC was 0.85. The Cronbach's α for the 5-item NFA was 0.83. A factor analysis was conducted on the two scales and confirmed to form two variables of individual difference. Next, we employed the same mean split (MNFC = 4.34; MNFA = 5.58) to divide the respondents into four groups and analysed the high-NFC/low-NFA (N = 32) and high-NFA/low-NFC (N = 64). As in Study 1, the Wald chi-square statistic and parameter estimates were measured using the choice-based multinomial logit model. Table 10.2 shows the results of this time-constrained DCE.

High-NFC Individuals' Preference

When respondents were under the time constraint, for the unfamiliar brand, the high-NFC respondents still preferred the localised informational elements (β-PN = 0.516, Wald-PN = 10.350, $p < 0.01$; β-PD = 0.390, Wald-PD = 6.911, $p < 0.01$). However, the standard BT influenced the choice of high-NFC respondents the most (β-BT = -0.701, Wald-BT = 29.020, $p < 0.001$). The same result was captured for the familiar brand, and it showed that the standard BT was the only attribute with significant influence (β-BT = -0.218, Wald-BT = 4.083, $p < 0.05$). These results suggest that the high-NFC individuals processed information using the peripheral route rather than the central route when the choice time was limited.

However, the Wald statistic clearly showed that the high-NFC respondents' preference for localised informational elements reduced. When these respondents faced time pressure, their preference for the localised PN and PD reduced by 44.3% and 61.3%, respectively, compared with their preference for the same packaging elements in the no time pressure condition. Further, their preference for the standard BT also dramatically increased (Wald-BT from 10.04 to 29.92; 190% increase), whereas there was no obvious change in their preference for the standard CE (Wald-CE from 0.32 to 0.27). Furthermore, for the familiar brand, the localised informational elements did not significantly influence the high-NFC

Table 10.2 Parameter estimates and chi-square of discrete choice experiment (with time constraint)

Subjects	*Brand familiarity*	*Attribute in localisation*	*Estimate*	*Chi-square*
High-NFC	Familiar brand	PN = Product name (L)	-3.773	0.000
		PD = Product description (L)	-0.041	0.101
		BT = Brand name and tagline (L)	-0.218***	4.083
		CC = Cultural cues (L)	0.041	0.101
High-NFC	Unfamiliar brand	PN = Product name (L)	0.516**	10.350
		PD = Product description (L)	0.390***	6.911
		BT = Brand name and tagline (L)	-0.701*	29.020
		CC = Cultural cues (L)	0.126	0.269
High-NFA	Familiar brand	PN = Product name (L)	0.132	1.143
		PD = Product description (L)	-0.073	0.437
		BT = Brand name and tagline (L)	-0.726*	68.470
		CC = Cultural cues (L)	0.155	1.729
High-NFA	Unfamiliar brand	PN = Product name (L)	-0.036	0.108
		PD = Product description (L)	0.231***	4.516
		BT = Brand name and tagline (L)	-0.590*	47.910
		CC = Cultural cues (L)	-0.009	0.000

Notes
L = Localised. * $p < 0.001$; ** $p < 0.01$; *** $p < 0.05$.

respondents' choice. Instead, they were more likely to prefer the standard BT (β-BT = -0.218, Wald-BT = 4.083, $p < 0.05$).

We also utilised DCE to generate the most effective localised packaging combinations of unfamiliar and familiar brands for high-NFC individuals. The results showed that for the unfamiliar brand, the combination of PN(L)** + PD(L)** + BT(S)* + CE(L) achieved the highest utility value = 1.732. In contrast, the most effective combination for the familiar brand was PN(L or S) + PD(S) + BT(S)*** + CE(L) with utility value = 0.3.

High-NFA Individuals' Preference

For both unfamiliar and familiar brands, the high-NFA respondents resorted to the peripheral route to process the information on the packaging. Their preference was significantly influenced by the standard BT (the familiar brand: β-BT = -0.726, Wald-BT = 68.470, $p < 0.001$; the unfamiliar brand: β-BT = -0.590, Wald-BT = 47.910, $p < 0.001$), while the localised PD of the unfamiliar brand also provided some impact (β-PD = 0.2321, Wald-BT = 4.516, $p < 0.05$).

As expected, for both unfamiliar and familiar brands, the perceived preference of the standard BT was enhanced when the choice time was controlled. The Wald statistic increased by 42.94%, compared with the same elements on the same packaging considered under the no time constraint condition. However, the high-NFA respondents did not show any preference for the standard CE, whereas they preferred the similar elements on the packaging when the choice time was available. The most effective localised packaging combinations for high-NFA individuals are as follows. For the unfamiliar brand, the combination is PN(S) + PD(L)*** + BT(S)* + CE(S) with utility value = 0.865. For the familiar brand, the combination is PN(L) + PD(S) + BT(S)* + CE(L) with utility value = 1.086.

Discussion

By extending Study 1, Study 2 shows that when resources are constrained, high-NFC and high-NFA individuals are both more likely to process information using the peripheral route. That is, the standard BT on the packaging becomes more effective in influencing consumers' choice, whereas the other symbolic cues, the cultural elements, do not contribute to the effectiveness of packaging localisation. Further, the resource availability is able to moderate the effectiveness of packaging elements in persuading consumers. Specifically, for the unfamiliar brand, the preference for localised informational elements decreased owing to the resource constraint, whereas the preference for some symbolic elements (i.e. standard BT) increased under the same condition.

The finding that the availability of choice time plays a key role in influencing persuasion effects has two theoretical contributions. First, it supports previous studies and confirms that the availability of choice time has a role in influencing consumers in processing information using a different route in the elaborate likelihood model (Haugtvedt et al., 1992). For example, the high-NFC individuals were still willing to evaluate the cognition-based packaging elements in localisation; however, because of time pressure, they were more likely to rely on standard affective-based packaging elements (resort to the peripheral route) to generate their choice behaviour. Second, our study demonstrates that time constraints may diminish the effectiveness of cognition-based messages (i.e. localised functional packaging elements) and enhance that of affective-based messages (i.e. standard symbolic packaging elements). This finding also supports the affective—cognitive model (Haddock et al., 2008). It confirmed that limited choice time can lead to affective responses from both high-NFC and high-NFA individuals. However, when there is no constraint on the choice time, a cognitive response is more likely to be elicited from high-NFC consumers, and the intensity of the affective response is diminished and replaced by the cognitive response following the increase in choice time.

11 Managers' Perspectives on Packaging Localisation

In-Depth Interviews with Brand Managers

For the purpose of this study, we identified five managers who manage their respective brands in China, and five other managers who similarly manage their respective brands in Pakistan (their identities are withheld for confidentiality reasons). The managers in China and Pakistan were responsible for various consumer goods, including ice cream, confectionary, snacks, carbonated soft drinks and detergents (see Table 11.1 for their demographics). All ten managers were interviewed telephonically. The primary objective of the interviews was to understand the reasons that major Western brands choose to localise their product packaging for China and Pakistan, and to compare their reasons with our findings from consumers' perspectives.

In general, localisation seems to be an umbrella strategy adopted by international brands in China and Pakistan. The brand managers consistently and unanimously believed that localisation is a crucial factor for gaining sales by enhancing their competitiveness in foreign markets, even with the full knowledge that they do not know whether consumers actually favour the localised packaging. They concurred that every brand has its own set of visual identity that makes it recognisable what the brand represents. Therefore, they adopted localisation in packaging to ensure consumers understand the information being communicated to them, especially when their first language is not English. Hence, they concluded that localisation is necessary. The managers specified that not all markets are prioritised for customised packaging using local languages as a cue, but China and Pakistan are large markets for international brands and thus they consider it worth investing in localisation. For example, a manager stated:

> Product/category relevance should be matched with brand name. However, owing to differences between different countries and culture, we realized the importance of localised package. The principle of localised package is similar to 'localised spokesperson'. In general, a global launched product uses different spokespersons in different countries. Similarly, a localised package may attract more local target consumers. A unified package conveys information of professionalisation, but a localised package indicates customisation.

DOI: 10.4324/9781003000822-13

Table 11.1 Managers' demographics

Respective countries	*Expertise*	*Type of product*	*Years of experience*
China	Consumer Insight Manager	FMCG products: Ice cream, chewing gum and snacks	5 months
China	Managing Director	Various FMCG products	20 years
China	Research Manager	Ice cream and snacks	14 years
China	Research Manager	Carbonated soft drink, potato chips, sanitary napkin, chocolates, cosmetic products, detergent	Undisclosed
Pakistan	Marketing Management	Beverages: Carbonated soft drinks, energy drink, juice	1 year
Pakistan	Product Management	Carbonated soft drinks	3 years
Pakistan	Brand Management	Shampoo	9 years
Pakistan	Sales Management	Food and beverages: Powdered milk, juices, cereals, coffee, confectionary	More than 3 years
Pakistan	Sales Management	Butter, soap, toothpaste, tea, ice cream, shampoo, cosmetic creams	4.5 years

The managers further revealed that the majority of brand sales derive from localised packaging, for the standard packaging is often available only at high-end or exclusive supermarkets. According to the Pakistan managers, the proportion of sales of localised to standard packaging was approximately 80:20 in Pakistan. The China managers did not offer any specific numbers but believed that the vast majority of sales are from localised packaging. The Pakistan managers also stated that the localised packaging is usually manufactured in Middle Eastern countries, such as Saudi Arabia or Dubai, and the supply chain is often shared amongst these channels for cost reduction purposes, since the Arabic and the Urdu language scripts are the same. The distribution is usually different owing to differences in target market needs and retailers' requirements. Therefore, the mass distribution of the localised packaging is found in most supermarkets, hypermarkets and grocery stores, whereas the distribution of the standard packaging is often limited to exclusive shopping centres or high-end shopping malls.

According to the managers in China, other than for regulatory compliance purposes, packaging localisation would ensure that consumers can recall their brands better and hence would like and buy the brands more often. To them, localisation can include using a native-language brand name as well as symbolic cues, such as pictures depicting the Chinese New Year. The managers in Pakistan do not have any legislative obligation to localise the packaging, but they similarly stated that the main purpose of localisation is to reach the mass audience. More importantly, they felt that customisation in terms of local language on packaging would give consumers a sense of belonging and a feeling that the company considers their needs. Manipulations, other than language, include the advertising of culturally symbolic cues on the packaging, which are often for special events or promoted as a limited edition. They exemplify this with a national flag on packaging on Independence Day, or religious symbols on festivals, such as Eid in Pakistan. For example, one manager remarked:

> Package localisation would be a consistent universal package design with a translated name in local language; sometimes, there might be more changes to adapt local culture or humanity habit. For example, the Coca-Cola has the same catchy red package universally which really helps you to find the product no matter where you are on this planet. How much effort does the manufacturer make on the package localisation shows the degree of respect to the local consumers?

Similarly, another manager outlined another reason pertaining to the importance of the local language:

> Localisation increases the consumers' acceptance of a product. I came across a study stating that 56.2% of consumers say that the ability to obtain information in their own language is more important than price. Here is the reference for the study: Common Sense Advisory, Can't Read, Won't Buy: Why Language Matters on Global Websites, 2006.

Some managers also cast doubts on whether standardisation would enhance the status of a product for most consumers. Another manager believed that there were possible negative effects of localising the packaging of foreign brands, and stated:

> Some manufacturers might think a localised product would downgrade the brand image with an imported background. However, even the premium brands have to lower their pride to get closer to consumers. Especially in the Chinese market, consumers with high purchasing power don't have high education level sometimes.

However, when asked whether they *objectively* know that Pakistani and Chinese consumers would prefer localised packaging, none of the managers was able to give a definite answer. The China managers admitted that this issue has not been examined earlier, and it is plausible that some products may be perceived to be better in standard packaging. They were also unsure how using a standard packaging would actually enhance quality perceptions or whether a price differential is possible for standard packaging. Interestingly, our findings suggest that packaging is the most important choice attribute, followed by price and then brand. This implies that Chinese consumers are willing to pay more for products in standard than in localised packaging, presumably because of the perceived foreignness and quality. A manager did admit that consumers might associate quality perceptions with standardisation, considering it to be imported while they may perceive localised packaging as local made. In relation to this point, the manager stated:

> Localised package can be more connected with and easily memorised by local people, and it can also help buzzing around crowds easier. But Chinese consumers tend to perceive the foreign brand as more premium and high quality; hence, it is possible that having a global package can help drive the brand premium and quality image.

Indeed, particularly given the recent safety and health concerns related to food products in China and Pakistan, products in standard packaging may be preferred owing to perceptions that the products are more likely to be imported and are of high quality and integrity. The managers agreed that standard packaging could change Pakistani and Chinese consumers' perceptions of the country of origin, and they might prefer standard packaging because of the value and quality perceptions created by the image of the origin country. They also pointed out that for certain hedonic products, such as chocolate, ice cream and cosmetic products, consumers might like the original name of the brand rather than a localised/translated brand name. Thus, it appears that although brand managers are cognisant of the *possibility* that standard packaging may enhance brand competitiveness, the strategy has not been examined objectively, let alone implemented. Consistent with this view, a manager mentioned: 'The original name might work better for an indulgent category, such as Häagen-Dazs. Also, premium skin care products usually keep the original name instead of a localised package'.

In summary, it seems that what international marketers currently practise is not fully aligned with the findings of our studies. For example, although the China manager admitted that Chinese consumers may perceive foreign brands in standard packaging as more premium and high quality, their primary objective is to mass-sell their products in China. Hence, standard (Western) packaging may backfire in that it may be viewed as targeting only certain premium segments, thereby leading to low sales and reduced market share. By contrast, localisation may blur the differences due to the country of origin and culture, thereby improving the mass appeal of the products to the 'average' Chinese consumers.

Following from the previous point, they questioned whether standard packaging would yield a price premium over localised packaging, even if standard packaging is preferred. Logically, if standard packaging leads to more favourable perceptions of quality and value (Marlow & Jansson-Boyd, 2011), it should follow that this packaging can support a higher price. According to the managers, they promote both standard and localised packaging at the same price. However, retailers often charge a premium on standard packaging, for such packaging is often not locally made, giving an impression of Westernisation.

In conclusion, it thus appears from these in-depth interviews that the managers unanimously acknowledged that they *did not know* whether local consumers actually preferred the localised packaging to the standard packaging. The managers also admitted that localising the brands' packaging might create multiple identities for the brand, consequently affecting its global identity or distinctiveness. Consistent with this argument, Zhang (2014) contends that packaging modifications reduce consumers' ability to recognise the brand, thereby leading to loss of sales opportunities. Those deciding the actual market practices are not particularly worried about research that shows that localisation may not necessarily work; or perhaps the confounding findings of advertising research do not apply to packaging, and hence, by conventional wisdom, *all* products should still be localised. Substantial knowledge gaps still remain about whether foreign product packaging should be localised and the conditions under which it should be localised. Apparently, no studies have addressed this question empirically, despite research establishing that packaging may be more important than advertising in influencing consumers, especially when packaging serves as the last point of communication with consumers before their actual purchase (Thomas, 2000). Thus, it seems that much more research is required in this area in order to fully disentangle the differential effects of different packaging cues, and to reconcile research findings with brand managers' practices.

Conclusion

That the centre of economic gravity has shifted East was a foregone fact by the start of the 20th century, and Asia remains the main engine of global growth well into the 21st century. As Asian countries develop, their middle-class population increases, and consumers become more affluent. Further, industries catering to these markets are also experiencing stellar growth over the past few decades, driven largely by the likes of China and India with their large domestic population. This fact and opportunity has not escaped the eyes of Western consumer packaged goods (CPG) brands, as they make exports to these markets a key pillar of their global growth strategy. The internationalisation of many brands within the CPG industry continues to increase, as brands seek out new high-growth sectors and markets in Asia.

As this book has demonstrated through the studies reported across various chapters, localising the packaging of Western CPG brands in Asia is an effective marketing strategy to target Asian consumers more effectively. Apparently, localising the packaging allows consumers to comprehend the products better. It also enables them to identify better with or associate themselves with the brands symbolically. Indeed, Nickels and Jolson (1976) stress the importance of packaging as a marketing strategy by designating it as the fifth P of the marketing mix, distinct from the product that it holds.

However, the approach is not as simple or straightforward as it seems, such as by translating the language on the packaging to the vehicular language of the target market. Different packaging elements (informational, graphic and structural) serve different functions, and there is more to packaging localisation than just translation, as exemplified by the studies in Part II regarding the differential influence of product types, consumption situations or even personality factors. In some instances, indiscriminately localising a packaging may even backfire because consumers express preference for the standard (i.e. original, non-localised) version of the packaging. For example, Chapter 7 shows that with hedonic products, standard packaging may actually work better than localised packaging in attracting consumers. Similarly, Chapter 8 demonstrates that consumers may be willing to pay a price premium for a conspicuously consumed product when the packaging is in its original (i.e. Western) form than one customised for the local market.

DOI: 10.4324/9781003000822-14

Thus, these collective findings suggest that packaging localisation is undoubtedly effective, but better care and scrutiny is needed to identify what, where and when product packaging should be localised because much more is not clearly understood regarding this important marketing communication tool. For instance, packaging is increasingly playing a multisensorial role in delivering the total experience that consumers have with a product, especially when they often do not, or cannot, see the product itself but only the external packaging that houses the product (Velasco & Spence, 2018). From a localisation perspective, an ensuing question is how it would be possible, through the packaging, to customise these multisensorial experiences in order to cater to local consumer taste or requirements. This issue is complicated by the fact that some products are consumed in, or used from, the packaging, such as toothpaste and insect sprays. In contrast, there are products such as instant noodles, where the packaging appears briefly and is discarded as the products are consumed. That is, the exposure time to the different types of packaging would necessarily require different strategic approaches in order to maximise the efficacy of packaging for marketing purposes.

Similarly, as with the general pivot towards green consumerism, such as through the use of refillable bottles, biodegradable wrapping or recycled paper, the issue of sustainability is another topic of interest in packaging localisation. Indeed, sustainability and ethical consumption is a developing theme among Asian CPG consumers because an increasing number of consumers are expressing the intention to buy sustainable products (Lewis, 2016; YouGov, 2021). These numbers ranged from as high as 79% for Indonesia to over half for Singapore and Hong Kong. Chinese consumers (65%) and Indian consumers (73%) have also indicated similar intentions to support products with sustainability characteristics (Lewis, 2016). It would be interesting to gain further understanding about the ways in which localisation from a sustainability perspective may serve as a competitive advantage for Western CPG brands in Asia.

Yet another area that needs further illumination concerns evolving global consumer trends. One specific area is the rapid growth in online retailing. Given that shopping online is an entirely different experience from shopping in a physical supermarket, the communicator role of packaging would also differ across shopping modes. Consumers would glance across dozens, or more, of competing brands seated in close proximity on physical shelves. In contrast, online shopping is limited by the screen size of the electronic device and may even be based on an elaborate search for particular products or brands. Hence, the visual elements of packaging and their impact are likely to function differently.

The emergence of domestic brands that have successfully competed against established international brands is yet another area of concern for Western CPG brands. Indeed, a recent report reveals that local brands in several Asian markets, including China, Indonesia, South Korea and Vietnam, are the category market leader in their respective countries (Kantar, 2019). This is largely due to local players' innate understanding of, and ability to adapt quickly to, local consumer trends. As the 21st-century consumers become less brand loyal, it gives rise to challenges for Western CPG brands in protecting their market share dominance. This situation further aligns with the need to comprehend how 'local' should Western CPG brands become to compete successfully in Asian markets. Perhaps, McKinsey & Company (2019) is not too far off when, in a recent report, the company suggests that the future of packaging lies in producing packaging that is not only smart and reusable, but also personalised to the requirements of the target audience. In this regard, one would expect that catering to local consumers in markets that are as culturally and geographically diverse as Asia would require Western CPG brands to firmly embrace the marketing strategy of localisation in order to remain relevant in the minds of Asian consumers.

References

Aaker, D. A. (1991). *Managing brand equity: Capitalizing on the value of a brand name.* New York, NY: Free Press.

Aaker, D. A. (1996). Measuring brand equity across products and markets. *California Management Review, 38*(3), 102–120.

Aaker, D. A. (2012). *Building strong brands.* New York, NY: Prometheus Global Media.

Accenture Consulting. (2018, 16 June). *A new era of trade for consumer goods industry: Why consumer goods companies need to think digital to reach traditional trade in Asia Pacific.* Accenture Consulting. www.accenture.com/_acnmedia/PDF-73/Accenture-new-era-of-trade-for-consumer-goods-industry.pdf

Ahn, J., & Ferle, C. L. (2008). Enhancing recall and recognition for brand names and body copy: A mixed-language approach. *Journal of Advertising, 37*(3), 107–117.

Ailawadi, K. L., Lehmann, D. R., & Neslin, S. A. (2003). Revenue premium as an outcome measure of brand equity. *Journal of Marketing, 67*(4), 1–17.

Alba, J. W., & Hutchinson, J. W. (1987). Dimensions of consumer expertise. *Journal of Consumer Research, 13*(4), 411–454.

Alcántara-Pilar, J. M., del Barrio-García, S., Crespo-Almendros, E., & Porcu, L. (2017). A review of psycho- vs. socio-linguistics theories: An application to marketing research. In I. R. M. Association (Ed.), *Advertising and branding: Concepts, methodologies, tools, and applications: Concepts, methodologies, tools, and applications* (1–30). Hershey, PA: IGI Global.

Alcántara-Pilar, J. M., del Barrio-García, S., Porcu, L., & Crespo-Almendros, E. (2017). Language as a cultural vessel in online servicescapes: Its impact on consumers' perceived risk, attitudes, and behavioural intentions. *Journal of Consumer Behaviour, 16*(6), 61–75. doi:10.1002/cb.1656

Alden, D. L., Steenkamp, J.-B. E., & Batra, R. (1999). Brand positioning through advertising in Asia, North America, and Europe: The role of global consumer culture. *Journal of Marketing, 63*(1), 75–87.

Amaldoss, W., & Jain, S. (2005). Pricing of conspicuous goods: A competitive analysis of social effects. *Journal of Marketing Research, 42*(1), 30–42.

Ampuero, O., & Vila, N. (2006). Consumer perceptions of product packaging. *Journal of Consumer Marketing, 23*(2), 100–112. https://doi.org/10.1108/07363760610655032

Angka, S., Hémar-Nicolas, V., Hapsari, H. P., & Olsen, A. (2020). How packaging colours and claims influence children's vegetable attitude and intake—An exploratory cross-cultural comparison between Indonesia and Denmark. *Food Quality and Preference, 79*, 103795.

Anselmsson, J., Bondesson, N. V., & Johansson, U. (2014). Brand image and customers' willingness to pay a price premium for food brands. *Journal of Product & Brand Management, 23*(2), 90–102.

Appel, M., Gnambs, T., & Maio, G. R. (2012). A short measure of the need for affect. *Journal of Personality Assessment, 94*(4), 418–426.

Appel, M., & Richter, T. (2010). Transportation and need for affect in narrative persuasion: A mediated moderation model. *Media Psychology, 13*(2), 101–135.

Aprile, M. C., Caputo, V., & Nayga, R. M. J. (2012). Consumers' valuation of food quality labels: The case of the European geographic indication and organic farming labels. *International Journal of Consumer Studies, 36*(2), 158–165.

Arboretti, R., & Bordignon, P. (2016). Consumer preferences in food packaging: CUB models and conjoint analysis. *British Food Journal, 118*(3), 527–540.

Ares, G., Besio, M., Giménez, A., & Deliza, R. (2010). Relationship between involvement and functional milk desserts intention to purchase. Influence on attitude towards packaging characteristics. *Appetite, 55*(2), 298–304.

Argo, J. J., Dahl, D. W., & Morales, A. C. (2006). Consumer contamination: How consumers react to products touched by others. *Journal of Marketing, 70*(2), 81–94.

Argo, J. J., & White, K. (2012). When do consumers eat more? The role of appearance self-esteem and food packaging cues. *Journal of Marketing, 76*(2), 67–80.

Aslam, M. M. (2006). Are you selling the right colour? A cross-cultural review of colour as a marketing cue. *Journal of Marketing Communications, 12*(1), 15–30.

Atsmon, Y., & Magni, M. (2012). *Meet the Chinese consumer of 2020.* www.mckinsey.com/global-themes/asia-pacific/meet-the-chinese-consumer-of-2020

Baccarella, C. V., Maier, L., & Voigt, K.-I. (2021). How consumption-supportive packaging functionality influences consumers' purchase intentions: The mediating role of perceived product meaningfulness. *European Journal of Marketing, 55*(8), 2239–2268.

Bai, H., McColl, J., & Moore, C. (2021). Luxury fashion retailers' localised marketing strategies in practice–evidence from China. *International Marketing Review.* https://doi.org/10.1108/IMR-02-2021-0079

Bakar, A., Khan, H., Hashim, N. H., & Lee, R. (2021). The strange bedfellows of packaging cues and religiosity. *Journal of Global Scholars of Marketing Science.* https://doi.org/10.1080/21639159.2021.1885302

Bakar, A., Lee, R., & Rungie, C. (2013). The effects of religious symbols in product packaging on Muslim consumer responses. *Australasian Marketing Journal, 21*(3), 198–204.

Baker, T., Hunt, J., & Scribner, L. (2002). The effect of introducing a new brand on consumer perceptions of current brand similarity: The roles of product knowledge and involvement. *Journal of Marketing Theory and Practice, 10*(4), 45–57.

Baker, W., Hutchinson, J., Moore, D., & Nedungadi, P. (1986). Brand familiarity and advertising: Effects on the evoked set and brand preference. In R. J. Lutz (Ed.), *Advances in consumer research* (Vol. 13, 637–642). Provo, UT: Association for Consumer Research.

Bamiatzi, V., Bozos, K., Cavusgil, S. T., & Hult, G. T. M. (2016). Revisiting the firm, industry, and country effects on profitability under recessionary and expansion periods: A multilevel analysis. *Strategic Management Journal, 37*(7), 1448–1471.

Baptista, I., Valentin, D., Saldaña, E., & Behrens, J. (2021). Effects of packaging color on expected flavor, texture, and liking of chocolate in Brazil and France. *International Journal of Gastronomy and Food Science, 24,* 100340. https://doi.org/10.1016/j.ijgfs.2021.100340

Baron, R. M., & Kenny, D. A. (1986). The moderator-mediator variable distinction in social psychological research: Conceptual, strategic, and statistical considerations. *Journal of Personality and Social Psychology, 51*(6), 1173–1182.

Barrera, G. A., & Ponce, H. R. (2021). Personality traits influencing young adults' conspicuous consumption. *International Journal of Consumer Studies, 45*(3), 335–349.

Bartels, M., Tillack, K., & Jordan Lin, C. T. (2018). Communicating nutrition information at the point of purchase: An eye-tracking study of shoppers at two grocery stores in the United States. *International Journal of Consumer Studies, 42*(5), 557–565. https://doi.org/10.1111/ijcs.12474

Bartsch, F., Riefler, P., & Diamantopoulos, A. (2016). A taxonomy and review of positive consumer dispositions toward foreign countries and globalization. *Journal of International Marketing, 24*(1), 82–110. doi:10.1509/jim.15.0021

Baruk, A. I., & Iwanicka, A. (2015). Polish final purchasers' expectations towards the features of dairy product packaging in the context of buying decisions. *British Food Journal, 117*(1), 178–194.

Batra, R., & Homer, P. M. (2004). The situational impact of brand image beliefs. *Journal of Consumer Psychology, 14*(3), 318–330.

Batra, R., Ramaswamy, V., Alden, D. L., Steenkamp, J. B. E., & Ramachander, S. (2000). Effects of brand local and nonlocal origin on consumer attitudes in developing countries. *Journal of Consumer Psychology, 9*(2), 83–95.

Bauer, H. H., Heinrich, D., & Schäfer, D. B. (2012). The effects of organic labels on global, local, and private brands: More hype than substance? *Journal of Business Research, 66*(8), 1035–1043. https://doi.org/10.1016/j.jbusres.2011.12.028

Beck, J. T., Chapman, K., & Palmatier, R. W. (2015). Understanding relationship marketing and loyalty program effectiveness in global markets. *Journal of International Marketing, 23*(3), 1–21.

Becker, L., van Rompay, T. J. L., Schifferstein, H. N. J., & Galetzka, M. (2011). Tough package, strong taste: The influence of packaging design on taste impressions and product evaluations. *Food Quality and Preference, 22*(1), 17–23. http://dx.doi.org/10.1016/j.foodqual.2010.06.007

Bell, D. R., Corsten, D., & Knox, G. (2011). From point of purchase to path to purchase: How preshopping factors drive unplanned buying. *Journal of Marketing, 75*(1), 31–45.

Berk, Z. (2009). *Food process engineering and technology.* San Diego, CA: Academic Press.

Bettman, J. R. (1979). Memory factors in consumer choice: A review. *Journal of Marketing, 43*(2), 37–53.

Bettman, J. R., & Sujan, M. (1987). Effects of framing on evaluation of comparable and noncomparable alternatives by expert and novice consumers. *Journal of Consumer Research, 26*(4), 141–154.

Bezaz, N., & Kacha, M. (2021). An experimental study of the effect of packaging colour on children's evaluation of packaging and attitude towards the brand. *International Journal of Retail & Distribution Management, 49*(6), 701–716. https://doi.org/10.1108/IJRDM-01-2020-0024

Bishop, M. M., Brocato, E. D., & Vijayalakshmi, A. (2015). The role of medium content and ad format congruity in influencing advertising outcomes. *Journal of Marketing Communications, 2*(1), 1–14.

Blatt, D. (2002). *POP measures up—Learning from the supermarket class of trade.* Washington, DC: Point of Purchase Advertising Institute.

Bloch, P. H. (1995). Seeking the ideal form: Product design and consumer response. *Journal of Marketing, 59*(July), 16–29.
Bloch, P. H. (2011). Product design and marketing: Reflections after fifteen years. *Journal of Product Innovation Management, 28*(3), 378–380.
Blommaert, J. (2010). *The sociolinguistics of globalization*. Cambridge: Cambridge University Press.
Blumer, H. (1969). Fashion: From class differentiation to collective selection. *The Sociological Quarterly, 10*(3), 275–291.
Bose, I. (2019, 10 August). *How small sizes are packing a punch in retail*. Financial Express. www.financialexpress.com/industry/how-small-sizes-are-packing-a-punch-in-retail/1658452/
Boyce, C., & Neale, P. (2006). *Conducting in-depth interviews: A guide for designing and conducting in-depth interviews for evaluation input*. Watertown, MA: Pathfinder International.
Breetz, C. (2014). *Product packaging as tool to demand a price premium: Does packaging enhance consumers 'value perception to justify a price premium*. Hamburg: Anchor Academic Publishing.
Brooks, C. (1994). *Sports marketing: Competitive business strategies for sports*. Englewood Cliffs, NJ: Prentice-Hall.
Brumbaugh, A. M. (2002). Source and nonsource cues in advertising and their effects on the activation of cultural and subcultural knowledge on the route to persuasion. *Journal of Consumer Research, 29*(2), 258–269.
Burbank, P. M., & Martins, D. C. (2010). Symbolic interactionism and critical perspective: Divergent or synergistic? *Nursing Philosophy, 11*(1), 25–41.
Busnaina, I., Almgatif, F., & Sdiri, I. (2021). Targeting arab markets: Is cultural context irrelevant to marketing in libya? *Management, 9*(3), 153–176.
Cacioppo, J. T., & Petty, R. E. (1982). The need for cognition. *Journal of Personality and Social Psychology, 42*(1), 116–131.
Cacioppo, J. T., Petty, R. E., & Feng Kao, C. (1984). The efficient assessment of need for cognition. *Journal of Personality Assessment, 48*(3), 306–307.
Çakır, M., Balagtas, J. V., Okrent, A. M., & Urbina-Ramirez, M. (2021). Effects of package size on household food purchases. *Applied Economic Perspectives and Policy, 43*(2), 781–801. https://doi.org/10.1093/aepp/ppz019
Campbell, D. T., & Fiske, D. W. (1959). Convergent and divergent validation by the multitraitmultimethod matrix. *Psychological Bulletin, 56*, 81–105.
Campbell, M. C., & Keller, K. L. (2003). Brand familiarity and advertising repetition effects. *Journal of Consumer Research, 30*(2), 292–304.
Celhay, F., Cheng, P., Masson, J., & Li, W. (2020). Package graphic design and communication across cultures: An investigation of Chinese consumers' interpretation of imported wine labels. *International Journal of Research in Marketing, 37*(1), 108–128. https://doi.org/10.1016/j.ijresmar.2019.07.004
Chandon, P. (2013). How package design and packaged-based marketing claims lead to overeating. *Applied Economic Perspectives and Policy, 35*(1), 7–31. https://doi.org/10.1093/aepp/pps028
Chandon, P., Hutchinson, J. W., Bradlow, E. T., & Young, S. H. (2009). Does in-store marketing work? Effects of the number and position of shelf facings on brand attention and evaluation at the point of purchase. *Journal of Marketing, 73*(6), 1–17.

Chang, C.-T., & Yen, C.-T. (2013). Missing ingredients in metaphor advertising: The right formula of metaphor type, product type, and need for cognition. *Journal of Advertising, 42*(1), 80–94. doi:10.1080/00913367.2012.749090

Chaudhuri, H. R., & Majumdar, S. (2006). Of diamonds and desires: Understanding conspicuous consumption from a contemporary marketing perspective. *Academy of Marketing Science Review, 11*(2/3), 2–18.

Chen, S.-S., Choubey, B., & Singh, V. (2021). A neural network based price sensitive recommender model to predict customer choices based on price effect. *Journal of Retailing and Consumer Services, 61.* https://doi.org/10.1016/j.jretconser.2021.102573

Chen, X., Barnes, C., Childs, T., Henson, B., & Shao, F. (2009). Materials' tactile testing and characterisation for consumer products' affective packaging design. *Materials & Design, 30*(10), 4299–4310.

Cheon, B. K., Melani, I., & Hong, Y.-Y. (2020). How USA-centric is psychology? An archival study of implicit assumptions of generalizability of findings to human nature based on origins of study samples. *Social Psychological and Personality Science, 11*(7), 928–937.

Cheon, H. J., Cho, C.-H., & Sutherland, J. (2007). A meta-analysis of studies on the determinants of standardization and localization of international marketing and advertising strategies. *Journal of International Consumer Marketing, 19*(4), 109–147.

Cheung, M.-C., Chan, A. S., & Sze, S. L. (2010). Electrophysiological correlates of brand names. *Neuroscience Letters, 485*(3), 178–182.

China Internet Watch. (2017, June). *Top 21 FMCG companies in China reaching over 100 million households in 2017, December, 2017.* China Internet Watch. www.chinainternetwatch.com/23024/fmcg-oct-2017/

Chiou, J. S., & Droge, C. (2015). The effects of standardization and trust on franchisee's performance and satisfaction: A study on franchise systems in the growth stage. *Journal of Small Business Management, 53*(1), 129–144.

Chitturi, R., Londoño, J. C., & Henriquez, M. C. (2021). Visual design elements of product packaging: Implications for consumers' emotions, perceptions of quality, and price. *Color Research and Application.* In press. https://doi.org/10.1002/col.22761

Chrysochou, P. (2017). *Consumer behavior research methods consumer perception of product risks and benefits* (409–428). Cham: Springer.

Chrysochou, P., & Festila, A. (2019). A content analysis of organic product package designs. *Journal of Consumer Marketing, 36*(4), 441–448. https://doi.org/10.1108/JCM-06-2018-2720

Chrysochou, P., & Grunert, K. G. (2014). Health-related ad information and health motivation effects on product evaluations. *Journal of Business Research, 67*(6), 1209–1217. http://dx.doi.org/10.1016/j.jbusres.2013.05.001

Chu, R., Tang, T., & Hetherington, M. M. (2021). The impact of food packaging on measured food intake: A systematic review of experimental, field and naturalistic studies. *Appetite, 166*, 105579. https://doi.org/10.1016/j.appet.2021.105579

Clement, J. (2007). Visual influence on in-store buying decisions: An eye-track experiment on the visual influence of packaging design. *Journal of Marketing Management, 23*(9–10), 917–928.

Clement, J., Kristensen, T., & Grønhaug, K. (2013). Understanding consumers' in-store visual perception: The influence of package design features on visual attention. *Journal of Retailing and Consumer Services, 20*(2), 234–239.

Clement, J., Smith, V., Zlatev, J., Gidlöf, K., & Van de Weijer, J. (2017). Assessing information on food packages. *European Journal of Marketing, 51*(1), 219–237.

Cleveland, M., & Chang, W. (2009). Migration and materialism: The roles of ethnic identity, religiosity, and generation. *Journal of Business Research, 62*(10), 963–971.

Cohen, A. R., Stotland, E., & Wolfe, D. M. (1955). An experimental investigation of need for cognition. *The Journal of Abnormal and Social Psychology, 51*(2), 291–294.

Cohen, E. (2009). Applying best-worst scaling to wine marketing. *International Journal of Wine Business Research, 21*(1), 8–23.

Cooley, C. H. (1902). *Human nature and the social order* (1st ed.). New York, NY: Charles Scribner's Sons.

Cooper, D. R., & Schindler, P. S. (2006). *Business research methods.* New York, NY: McGraw-Hill.

Craig, C. S., & Douglas, S. P. (2005). *International marketing research* (3rd ed.). Chichester, NJ: John Wiley & Sons.

Creusen, M. E., & Schoormans, J. P. (2005). The different roles of product appearance in consumer choice. *Journal of Product Innovation Management, 22*(1), 63–81.

Creusen, M. E. H., Gemser, G., & Candi, M. (2018). The influence of experiential augmentation on product evaluation. *European Journal of Marketing, 52*(5/6), 925–945. doi:10.1108/EJM-04-2016-0220

Cui, G., & Liu, Q. (2001). Emerging market segments in a transitional economy: A study of urban consumers in China. *Journal of International Marketing, 9*(1), 84–106.

Cumming, G. (2009). Inference by eye: Reading the overlap of independent confidence intervals. *Statistics in Medicine, 28*(2), 205–220.

Dawar, N., & Parker, P. (1994). Marketing universals: Consumers' use of brand name, price, physical appearance, and retailer reputation as signals of product quality. *Journal of Marketing, 58*(2), 81–95.

daxueconsulting. (2021, 1 November). *The importance of getting the size of your product right.* Daxue Group. https://daxueconsulting.com/packaging-sizes-china/

De Mooij, M. (2010). *Consumer behavior and culture: Consequences for global marketing and advertising.* Thousand Oaks, CA: Sage.

De Run, E. C. (2005). Does targeting really work? The perspective of a dominant ethnic group. *International Journal of Business and Society, 6*(1), 27–51.

de Sousa, M. M., Carvalho, F. M., & Pereira, R. G. (2020). Colour and shape of design elements of the packaging labels influence consumer expectations and hedonic judgments of specialty coffee. *Food Quality and Preference, 83*, 103902.

Decré, G. B., & Cloonan, C. (2019). A touch of gloss: Haptic perception of packaging and consumers' reactions. *Journal of Product & Brand Management, 28*(1), 117–132. https://doi.org/10.1108/JPBM-05-2017-1472

Delgado-Ballester, E., & Fernandez Sabiote, E. (2015). Brand experimental value versus brand functional value: Which matters more for the brand? *European Journal of Marketing, 49*(11/12), 1857–1879.

Delgado-Ballester, E., Navarro, A., & Sicilia, M. (2012). Revitalising brands through communication messages: The role of brand familiarity. *European Journal of Marketing, 46*(1), 31–51.

Deng, X., & Srinivasan, R. (2013). When do transparent packages increase (or decrease) food consumption? *Journal of Marketing, 77*(4), 104–117. https://doi.org/10.1509/jm.11.0610

Deshpandé, R., & Stayman, D. M. (1994). A tale of two cities: Distinctiveness theory and advertising effectiveness. *Journal of Marketing Research, 31*(1), 57–64.

Dhar, R., & Wertenbroch, K. (2000). Consumer choice between hedonic and utilitarian goods. *Journal of Marketing Research, 37*(1), 60–71.

Di Cicco, F., Zhao, Y., Wijntjes, M. W., Pont, S. C., & Schifferstein, H. N. (2021). A juicy orange makes for a tastier juice: The neglected role of visual material perception in packaging design. *Food Quality and Preference, 88*, 104086. https://doi.org/10.1016/j.foodqual.2020.104086

Diamantopoulos, A., Arslanagic-Kalajdzic, M., & Moschik, N. (2020). Are consumers' minds or hearts guiding country of origin effects? Conditioning roles of need for cognition and need for affect. *Journal of Business Research, 108*, 487–495.

Dickson, P. R., & Sawyer, A. G. (1990). The price, knowledge and search of supermarket shoppers. *Journal of Marketing, 54*(3), 42–53.

Dittmar, H. (1992). *The social psychology of material possessions: To have is to be.* New York, NY: Harvester Wheatsheaf.

Dobbs, R., Remes, J., Manyika, J., Roxburgh, C., Smit, S., & Schaer, F. (2012). *Urban world: Cities and the rise of the consuming class.* Washington, DC: McKinsey Global Institute.

Doğan, V., & Özkara, B. Y. (2013). Investigation of brand name-country of origin preference in four different product groups with respect to conspicuous consumption tendency. *International Review of Management and Marketing, 3*(4), 190–203.

Dörnyei, K. R. (2020). Limited edition packaging: Objectives, implementations and related marketing mix decisions of a scarcity product tactic. *Journal of Consumer Marketing, 37*(6), 617–627.

Drolet, A., Williams, P., & Lau-Gesk, L. (2007). Age-related differences in responses to affective vs. rational ads for hedonic vs. utilitarian products. *Marketing Letters, 18*(4), 211–221.

Dublish, S., Seaton, F., & Laskey, H. (2015). Effects of Spanish vs. English ads on bilingual Hispanics: Moderating role of language dominance. *Proceedings of the 1998 Academy of Marketing Science (AMS) annual conference.*

Dugan, R. G., Clarkson, J. J., & Beck, J. T. (2021). When cause-marketing backfires: Differential effects of one-for-one promotions on hedonic and utilitarian products. *Journal of Consumer Psychology, 31*(3). https://doi.org/10.1002/jcpy.1229

Durvasula, S., & Lysonski, S. (2010). Money, money, money—How do attitudes toward money impact vanity and materialism?—The case of young Chinese consumers. *Journal of Consumer Marketing, 27*(2), 169–179.

Eastman, J. K., & Eastman, K. L. (2015). Conceptualizing a model of status consumption theory: An exploration of the antecedents and consequences of the motivation to consume for status. *Marketing Management Journal, 25*(1), 1–15.

Eastman, J. K., Iyer, R., Eastman, K. L., Gordon-Wilson, S., & Modi, P. (2021). Reaching the price conscious consumer: The impact of personality, generational cohort and social media use. *Journal of Consumer Behaviour, 20*(4), 898–912. https://doi.org/10.1002/cb.1906

Eckhardt, G. M., Belk, R. W., & Wilson, J. A. (2015). The rise of inconspicuous consumption. *Journal of Marketing Management, 31*(7–8), 807–826.

Ekinci, Y., Sirakaya-Turk, E., & Preciado, S. (2013). Symbolic consumption of tourism destination brands. *Journal of Business Research, 66*(6), 711–718.

El Banna, A., Papadopoulos, N., Murphy, S. A., Rod, M., & Rojas-Méndez, J. I. (2018). Ethnic identity, consumer ethnocentrism, and purchase intentions among bi-cultural ethnic consumers:"Divided loyalties" or "dual allegiance"? *Journal of Business Research, 82*, 310–319.

Eldesouky, A., Pulido, A., & Mesias, F. (2015). The role of packaging and presentation format in consumers' preferences for food: An application of projective techniques. *Journal of Sensory Studies, 30*(5), 360–369.

Esch, F.-R., Langner, T., Schmitt, B., & Geus, P. (2006). Are brands forever? How brand knowledge and relationships affect current and future purchases. *Journal of Product & Brand Management, 15*(2), 98–105.

Esch, F.-R., Möll, T., Schmitt, B., Elger, C. E., Neuhaus, C., & Weber, B. (2012). Brands on the brain: Do consumers use declarative information or experienced emotions to evaluate brands? *Journal of Consumer Psychology, 22*(1), 75–85.

Espejel, J., Fandos, C., & Flavián, C. (2009). The influence of consumer involvement on quality signals perception: An empirical investigation in the food sector. *British Food Journal, 111*(11), 1212–1236.

Euromonitor. (2013). *World consumer income and expenditure patterns 2013*. Euromonitor. www.euromonitor.com/medialibrary/PDF/Book_WCIEP0.pdf

Euromonitor. (2015). *Report on packaged foods in Pakistan*. Euromonitor. www.euromonitor.com/packaged-food-in-pakistan/report

Euromonitor. (2018). *Packaged food in Pakistan*. Euromonitor. www.euromonitor.com/packaged-food-in-pakistan/report

Farrell, D., Gersch, U. A., & Stephenson, E. (2006). The value of China's emerging middle class. *McKinsey Quarterly, 2*(I), 60–69.

Fazio, R. H. (1990). Multiple processes by which attitudes guide behavior: The MODE model as an integrative framework. *Advances in Experimental Social Psychology, 23*, 75–109.

Feldman, J. M., & Lynch, J. G. (1988). Self-generated validity and other effects of measurement on belief, attitude, intention, and behavior. *Journal of Applied Psychology, 73*(3), 421–435.

Feng, A., & Adamson, B. (2018). Language policies and sociolinguistic domains in the context of minority groups in China. *Journal of Multilingual and Multicultural Development, 39*(2), 169–180.

Fenko, A., Lotterman, H., & Galetzka, M. (2016). What's in a name? The effects of sound symbolism and package shape on consumer responses to food products. *Food Quality and Preference, 51*, 100–108. https://doi.org/10.1016/j.foodqual.2016.02.021

Festila, A., & Chrysochou, P. (2018). Implicit communication of food product healthfulness through package design: A content analysis. *Journal of Consumer Behaviour, 17*(5), 461–476. https://doi.org/10.1002/cb.1732

Festila, A., Chrysochou, P., Hieke, S., & Massri, C. (2021). Public sensemaking of active packaging technologies: A feature-based perspective. *Public Understanding of Science*, 09636625211015830. https://doi.org/10.1177/09636625211015830

Fitzsimons, G. J., Hutchinson, J. W., Williams, P., Alba, J. W., Chartrand, T. L., Huber, J., . . . Russo, J. E. (2002). Non-conscious influences on consumer choice. *Marketing Letters, 13*(3), 269–279.

Folkes, V., & Matta, S. (2004). The effect of package shape on consumers' judgments of product volume: Attention as a mental contaminant. *Journal of Consumer Research, 31*(2), 390–401.

Ford, N., Trott, P., & Simms, C. (2016). Exploring the impact of packaging interactions on quality of life among older consumers. *Journal of Marketing Management, 32*(3–4), 275–312. https://doi.org/10.1080/0267257X.2015.1123758

Fornell, C., & Larcker, D. F. (1981). Evaluating structural equation models with unobservable variables and measurement error. *Journal of Marketing Research, 18*(1), 39–50.

Franco-Arellano, B., Vanderlee, L., Ahmed, M., Oh, A., & L'Abbé, M. (2020). Influence of front-of-pack labelling and regulated nutrition claims on consumers' perceptions of product healthfulness and purchase intentions: A randomized controlled trial. *Appetite, 149*, 104629.

Gao, H., Winterich, K. P., & Zhang, Y. (2016). All that glitters is not gold: How others' status influences the effect of power distance belief on status consumption. *Journal of Consumer Research, 43*(2), 265–281.

Garber, L. L. (1995). The package appearance in choice. *Association in Consumer Research, 22*, 653–660.

Garrido-Morgado, Á., González-Benito, Ó., Martos-Partal, M., & Campo, K. (2021). Which products are more responsive to in-store displays: Utilitarian or hedonic? *Journal of Retailing, 97*(3), 477–491.

Gelb, B. D. (2002). Observations: 'Market patriotism'—Advertising dilemma. *Journal of Advertising Research, 42*(1), 67–69.

Gerritsen, M., & Nickerson, C. (2010). Cross-cultural brand management and language choice. Response to the use of English in product advertisements in non-English speaking countries in Western and Southern Europe. *Paper presented at the 3rd Annual EuroMed Conference of the EuroMed Academy of Business.*

Gerritsen, M., Nickerson, C., Van Hooft, A., Van Meurs, F., Nederstigt, U., Starren, M., & Crijns, R. (2007). English in product advertisements in Belgium, France, Germany, the Netherlands and Spain. *World Englishes, 26*(3), 291–315.

Gierl, H., & Huettl, V. (2010). Are scarce products always more attractive? The interaction of different types of scarcity signals with products' suitability for conspicuous consumption. *International Journal of Research in Marketing, 27*(3), 225–235.

Gil-Pérez, I., Rebollar, R., & Lidón, I. (2020). Without words: The effects of packaging imagery on consumer perception and response. *Current Opinion in Food Science, 33*, 69–77. https://doi.org/10.1016/j.cofs.2019.12.006

Glaser, B. G. (2008). *Doing quantitative grounded theory.* Mill Valley, CA: Sociology Press.

Gofman, A., Moskowitz, H. R., Fyrbjork, J., Moskowitz, D., & Mets, T. (2009). Extending rule developing experimentation to perception of food packages with eye tracking. *Open Food Science Journal, 3*, 66–78. https://doi.org/10.2174/1874256400903010066

Green, C. L. (1999). Ethnic evaluations of advertising: Interaction effects of strength of ethnic identification, media placement, and degree of racial composition. *Journal of Advertising, 28*(1), 49–64.

Griskevicius, V., Tybur, J. M., Sundie, J. M., Cialdini, R. B., Miller, G. F., & Kenrick, D. T. (2007). Blatant benevolence and conspicuous consumption: When romantic motives elicit strategic costly signals. *Journal of Personality and Social Psychology, 93*(1), 85.

Grunert, K. G., & Wills, J. M. (2007). A review of European research on consumer response to nutrition information on food labels. *Journal of Public Health, 15*(5), 385–399. https://doi.org/10.1007/s10389-007-0101-9

Guo, X. (2013). Living in a global world: Influence of consumer global orientation on attitudes toward global brands from developed versus emerging countries. *Journal of International Marketing, 21*(1), 1–22.

Gupta, S., & Wright, O. (2019, February). *How global brands can respond to local competitors?* https://hbr.org/2019/02/how-global-brands-can-respond-to-local-competitors

Haddock, G., Maio, G. R., Arnold, K., & Huskinson, T. (2008). Should persuasion be affective or cognitive? The moderating effects of need for affect and need for cognition. *Personality and Social Psychology Bulletin, 34*(6), 769–778.
Hagtvedt, H., & Patrick, V. M. (2014). Consumer response to overstyling: Balancing aesthetics and functionality in product design. *Psychology & Marketing, 31*(7), 518–525.
Håkansson, N., Kvarnström, E., & Nilsson, E. (2014). *Using the package to influence consumers' choice of brand: A study on the effect of package communication claims on propensity to switch brand* (Masters). Uppsala University, Disciplinary Domain of Humanities and Social Sciences, Faculty of Social Sciences, Department of Business Studies, Sweden.
Han, Y. J., Nunes, J. C., & Drèze, X. (2010). Signaling status with luxury goods: The role of brand prominence. *Journal of Marketing, 74*(4), 15–30.
Hao, Y., Liu, H., Chen, H., Sha, Y., Ji, H., & Fan, J. (2019). What affect consumers' willingness to pay for green packaging? Evidence from China. *Resources, Conservation and Recycling, 141*, 21–29. https://doi.org/10.1016/j.resconrec.2018.10.001
Haugtvedt, C. P., Petty, R. E., & Cacioppo, J. T. (1992). Need for cognition and advertising: Understanding the role of personality variables in consumer behavior. *Journal of Consumer Psychology, 1*(3), 239–260.
He, J., & Wang, C. L. (2017). How global brands incorporating local cultural elements increase consumer purchase likelihood: An empirical study in China. *International Marketing Review, 34*(4), 463–479. https://doi.org/10.1108/IMR-08-2014-0272
Heinberg, M., Ozkaya, H. E., & Taube, M. (2016). A brand built on sand: Is acquiring a local brand in an emerging market an ill-advised strategy for foreign companies? *Journal of the Academy of Marketing Science, 44*(5), 586–607.
Heine, S. J., Lehman, D. R., Peng, K., & Greenholtz, J. (2002). What's wrong with cross-cultural comparisons of subjective Likert scales?: The reference-group effect. *Journal of Personality and Social Psychology, 82*(6), 903. https://doi.org/10.1037/0022-3514.82.6.903
Henderson, P. W., & Cote, J. A. (1998). Guidelines for selecting or modifying logos. *Journal of Marketing, 62*(2), 14–30.
Henderson, P. W., Giese, J. L., & Cote, J. A. (2004). Impression management using typeface design. *Journal of Marketing, 68*(4), 60–72. https://doi.org/10.1509/jmkg.68.4.60.42736
Henrich, J., Heine, S. J., & Norenzayan, A. (2010a). Most people are not WEIRD. *Nature, 466*(7302), 29–29.
Henrich, J., Heine, S. J., & Norenzayan, A. (2010b). The weirdest people in the world? *Behavioral and Brain Sciences, 33*(2–3), 61–83.
Herbes, C., Beuthner, C., & Ramme, I. (2020). How green is your packaging—A comparative international study of cues consumers use to recognize environmentally friendly packaging. *International Journal of Consumer Studies, 44*(3), 258–271.
Herbig, P. A. (1998). *Handbook of cross-cultural marketing*. New York, NY: The Haworth Press.
Hill, C. W. L., & Hult, G. T. M. (2016). *International business: Competing in the global marketplace*. New York, NY: McGraw-Hill Education.
Hilton, J. L., & Darley, J. M. (1991). The effects of interaction goals on person perception. In M. P. Zanna (Ed.), *Advances in experimental social psychology* (Vol. 24, 235–267). New York, NY: Academic Press.
Hine, T. (1995). *The total package: The secret history and hidden meanings of boxes, bottles, cans and other persuasive containers*. New York, NY: Back Bay Books.

Hirschman, E. C., & Holbrook, M. B. (1982). Hedonic consumption: Emerging concepts, methods and propositions. *Journal of Marketing, 46*(3), 92–101.

Ho, H.-C., Chiu, C. L., Jiang, D., Shen, J., & Xu, H. (2019). Influence of language of packaging labels on consumers' buying preferences. *Journal of Food Products Marketing*, 1–27. doi:10.1080/10454446.2019.1572562

Ho, J., Poh, F., Zhou, J., & Zipster, D. (2019, 15 March). *China consumer report 2020: The many faces of the Chinese consumer.* McKinsey and Company. www.mckinsey.com/featured-insights/china/china-consumer-report-2020-the-many-faces-of-the-chinese-consumer#

Hofstede, G. (2001). *Culture's consequences: Comparing values, behaviors, institutions and organizations across nations.* Beverly Hills, CA: Sage Publications.

Hoijer, H. (1954). The Sapir-Whorf hypothesis. In B. G. Blount (Ed.), *Language, culture and society: A book of readings* (120–131). Cambridge, MA: Winthrop.

Holmquist, J. P., & Cudmore, B. A. (2013). English in Korean advertising: An exploratory study. *International Journal of Marketing Studies, 5*(3), 94–103.

Homburg, C., Schwemmle, M., & Kuehnl, C. (2015). New product design: Concept, measurement, and consequences. *Journal of Marketing, 79*(3), 41–56.

Homer, P. M. (2006). Relationships among ad-induced affect, beliefs, and attitudes: Another look. *Journal of Advertising, 35*(1), 35–51.

Hong, S.-T., & Wyer, R. S. J. (1989). Effects of country-of-origin and product-attribute information on product evaluation: An information processing perspective. *Journal of Consumer Research, 16*(2), 175–187.

Hornikx, J., & van Meurs, F. (2021). *Foreign languages in advertising: Linguistic and marketing perspectives.* Cham: Palgrave MacMillan.

Hornikx, J., van Meurs, F., & Starren, M. (2007). An empirical study of readers' associations with multilingual advertising: The case of French, German and Spanish in Dutch advertising. *Journal of Multilingual and Multicultural Development, 28*(3), 204–219.

Hoskisson, R. E., Eden, L., Lau, C. M., & Wright, M. (2000). Strategy in emerging economies. *Academy of Management Journal, 43*(3), 249–267.

Hoyer, W. D., & Brown, S. P. (1990). Effects of brand awareness on choice for a common, repeat-purchase product. *Journal of Consumer Research, 17*(2), 141–148.

Huber, J., & Zwerina, K. (1996). The importance of utility balance in efficient choice designs. *Journal of Marketing Research, 33*(3), 307–317.

Hubert, M., Hubert, M., Florack, A., Linzmajer, M., & Kenning, P. (2013). Neural correlates of impulsive buying tendencies during perception of product packaging. *Psychology & Marketing, 30*(10), 861–873. https://doi.org/10.1002/mar.20651

Hunt, E., & Agnoli, F. (1991). The Whorfian hypothesis: A cognitive psychology perspective. *Psychological Review, 98*(3), 377–389.

Husić-Mehmedović, M., Omeragić, I., Batagelj, Z., & Kolar, T. (2017). Seeing is not necessarily liking: Advancing research on package design with eye-tracking. *Journal of Business Research, 80*, 145–154. https://doi.org/10.1016/j.jbusres.2017.04.019

Huskinson, T. L., & Haddock, G. (2004). Individual differences in attitude structure: Variance in the chronic reliance on affective and cognitive information. *Journal of Experimental Social Psychology, 40*(1), 82–90.

Ikonen, I., Sotgiu, F., Aydinli, A., & Verlegh, P. W. (2020). Consumer effects of front-of-package nutrition labeling: An interdisciplinary meta-analysis. *Journal of the Academy of Marketing Science, 48*(3), 360–383.

Ilich, K. L., & Hardey, M. (2020). 'It's all about the packaging': Investigation of the motivations, intentions, and marketing implications of sharing photographs of secondary packaging on Instagram. *Information, Communication & Society, 23*(1), 1–19. https://doi.org/10.1080/1369118X.2018.1478983

Inman, J. J., Winer, R. S., & Ferraro, R. (2009). The interplay among category characteristics, customer characteristics, and customer activities on in-store decision making. *Journal of Marketing, 73*(5), 19–29.

Jacobs, L., Keown, C., Worthley, R., & Ghymn, K. I. (1991). Cross-cultural colour comparisons: Global marketers beware! *International Marketing Review, 8*(3), 21–30.

Jo, M.-S., Nakamoto, K., & Nelson, J. E. (2003). The shielding effects of brand image against lower quality countries-of-origin in global manufacturing. *Journal of Business Research, 56*(8), 637–646.

Kacen, J. J., Hess, J. D., & Walker, D. (2012). Spontaneous selection: The influence of product and retailing factors on consumer impulse purchases. *Journal of Retailing and Consumer Services, 19*(6), 578–588.

Kahn, B. E., Deng, X., & Krishna, A. (2010). *Effects on visual weight perceptions of product image locations on packaging.* New York, NY: Routledge.

Kantar. (2019, 2 November). *ASIA BRAND FOOTPRINT 2019. Building a sustainable brand: A ranking of the most chosen consumer brands.* Kantar. www.kantar.com/inspiration/fmcg/asias-most-chosen-fmcg-brands-ranking

Kapferer, J. N. (1995). Brand cofusion—Empirical study of a legal concept. *Psychology & Marketing, 12*(6), 551–568.

Karnal, N., Machiels, C. J. A., Orth, U. R., & Mai, R. (2016). Healthy by design, but only when in focus: Communicating non-verbal health cues through symbolic meaning in packaging. *Food Quality and Preference, 52*, 106–119. http://dx.doi.org/10.1016/j.foodqual.2016.04.004

Kauppinen-Räisänen, H., & Luomala, H. T. (2010). Exploring consumers' product-specific colour meanings. *Qualitative Market Research: An International Journal, 13*(3), 287–308. https://doi.org/10.1108/13522751011053644

Kazmi, K. H. (2019, October). *Affection of consumers buying behavior in Pakistan.* www.pakistaneconomist.com/2019/10/07/affection-of-consumers-buying-behavior-in-pakistan/

Kelle, U. (2010). *The development of categories: Different approaches in grounded theory.* Los Angeles, CA: Cambridge University Press.

Keller, K. L. (1993a). Conceptualizing, measuring, and managing customer-based brand equity. *Journal of Marketing, 57*(1), 1–22.

Keller, K. L. (1993b). Memory retrieval factors and advertising effectiveness. In A. Mitchell (Ed), *Advertising exposure, memory, and choice* (11–48). Hillsdale, NJ: Lawrence Erlbaum Associates.

Keller, K. L. (1998). *Strategic brand management.* Upper Saddle River, NJ: Prentice-Hall.

Keller, K. L. (2003). Brand synthesis: The multidimensionality of brand knowledge. *Journal of Consumer Research, 29*(4), 595–600.

Keller, K. L., & Staelin, R. (1987). Effects of quality and quantity of information on decision effectiveness. *Journal of Consumer Research, 14*(2), 200–213.

Kent, R. J., & Allen, C. T. (1994). Competitive interference effects in consumer memory for advertising: The role of brand familiarity. *The Journal of Marketing, 58*(3), 97–105.

Ketelsen, M., Janssen, M., & Hamm, U. (2020). Consumers' response to environmentally-friendly food packaging—A systematic review. *Journal of Cleaner Production, 254*, 120123. https://doi.org/10.1016/j.jclepro.2020.120123

Khan, H. (2019). Effects of personal dispositions, familiarity and consumption situation on Western brands' packaging. *Journal of Consumer Marketing, 36*(6), 715–727.

Khan, H., & Bamber, D. (2007). Market entry using country-of-origin intelligence in an emerging market. *Journal of Small Business and Enterprise Development, 14*(1), 22–35.

Khan, H., & Lee, R. (2020a). A sociolinguistic perspective of the effects of packaging in bilingual markets. *Journal of Brand Management, 27*(2), 130–142.

Khan, H., & Lee, R. (2020b). Does packaging influence taste and quality perceptions across varying consumer demographics? *Food Quality and Preference*. doi:10.1016/j.foodqual.2020.103932

Khan, H., Lee, R., & Lockshin, L. (2015). Localising the packaging of foreign food brands: A case of Muslim consumers in Pakistan. *Journal of Product & Brand Management, 24*(4), 386–398.

Khan, H., Lee, R., & Lockshin, L. (2017). The effects of packaging localisation of Western brands in non-Western emerging markets. *Journal of Product & Brand Management, 26*, 589–599.

Khan, H., Lockshin, L., Lee, R., & Corsi, A. (2017). When is it necessary to localise product packaging? *Journal of Consumer Marketing, 34*(5), 373–383.

King, D., & Auschaitrakul, S. (2021). Affect-based nonconscious signaling: When do consumers prefer negative branding? *Psychology & Marketing, 38*(2), 338–358.

Kirmani, A., & Baumgartner, H. (2000). Reference points used in quality and value judgements. *Marketing Letters, 11*(4), 299–310.

Klimchuk, M. R., & Krasovec, S. A. (2012). *Packaging design: Successful product branding from concept to shelf* (2nd ed.). Chichester, NJ: John Wiley & Sons.

Kniazeva, M., & Belk, R. W. (2007). Packaging as vehicle for mythologizing the brand. *Consumption Markets & Culture, 10*(1), 51–69. https://doi.org/10.1080/10253860601164627

Köhler, W. (1967). Gestalt psychology. *Psychologische Forschung, 31*(1), XVIII–XXX.

Konstantoglou, A., Folinas, D., & Fotiadis, T. (2021). Comparison of consumers and industry managers concerning food packaging elements. *British Food Journal, 123*(3), 1103–1120. https://doi.org/10.1108/BFJ-04-2020-0295

Koo, J., & Suk, K. (2016). The effect of package shape on calorie estimation. *International Journal of Research in Marketing, 33*(4), 856–867. https://doi.org/10.1016/j.ijresmar.2016.03.002

Koslow, S., Shamdasani, P. N., & Touchstone, E. E. (1994). Exploring language effects in ethnic advertising: A sociolinguistic perspective. *Journal of Consumer Research, 20*(4), 575–585.

Kotler, P., & Keller, K. L. (2016). *Marketing management* (15th ed.). Upper Saddle River, NJ: Pearson Education Inc.

Kozinets, R. V. (2019). *Netnography: The essential guide to qualitative social media research*. London: Sage Publications Ltd.

Kravets, O., & Sandikci, O. (2014). Competently ordinary: New middle class consumers in the emerging markets. *Journal of Marketing, 78*(4), 125–140.

Krishna, A., & Ahluwalia, R. (2008). Language choice in advertising to bilinguals: Asymmetric effects for multinationals versus local firms. *Journal of Consumer Research, 35*(4), 692–705.

Krishna, A., Cian, L., & Aydınoğlu, N. Z. (2017). Sensory aspects of package design. *Journal of Retailing, 93*(1), 43–54.

Krishna, A., & Morrin, M. (2008). Does touch affect taste? The perceptual transfer of product container haptic cues. *Journal of Consumer Research, 34*(6), 807–818.

Krueger, R. A., & Casey, M. A. (2014). *Focus groups: A practical guide for applied research* (5th ed.). Thousand Oaks, CA: Sage Publications.

Kühn, S., Strelow, E., & Gallinat, J. (2016). Multiple "buy buttons" in the brain: Forecasting chocolate sales at point-of-sale based on functional brain activation using fMRI. *NeuroImage, 136*, 122–128. https://doi.org/10.1016/j.neuroimage.2016.05.021

Kulviwat, S., Bruner II, G. C., & Al-Shuridah, O. (2009). The role of social influence on adoption of high tech innovations: The moderating effect of public/private consumption. *Journal of Business Research, 62*(7), 706–712.

Labrecque, L. I., & Milne, G. R. (2012). Exciting red and competent blue: The importance of color in marketing. *Journal of the Academy of Marketing Science, 40*(5), 711–727.

Laroche, M., Pons, F., & Richard, M.-O. (2009). The role of language in ethnic identity measurement: A multitrait-multimethod approach to construct validation. *Journal of Social Psychology, 149*(4), 513–540.

Lau, H. T., & Lee, R. (2018). Ethnic media advertising effectiveness, influences and implications. *Australasian Marketing Journal, 26*(3), 216–220. https://doi.org/10.1016/j.ausmj.2018.05.014

Le Page, R. B. (1997). The evolution of a sociolinguistic theory of language. In F. Coulmas (Ed.), *The handbook of sociolinguistics* (374–385). Oxford: Blackwell Publishers Inc.

Leclerc, F., Schmitt, B. H., & Dubé, L. (1994). Foreign branding and its effects on product perceptions and attitudes. *Journal of Marketing Research, 31*(2), 263–270.

Lee, J. S. (2006). Linguistic constructions of modernity: English mixing in Korean television commercials. *Language in Society, 35*(1), 59–91.

Lee, R., Lee, K.-T., & Li, J. Y. (2017). A memory theory perspective of consumer ethnocentrism and animosity. *European Journal of Marketing, 51*(7/8), 1266–1285.

Lee, W.-C. J., Shimizu, M., Kniffin, K. M., & Wansink, B. (2013). You taste what you see: Do organic labels bias taste perceptions? *Food Quality and Preference, 29*(1), 33–39. http://dx.doi.org/10.1016/j.foodqual.2013.01.010

Lessig, V. P., & Park, C. W. (1982). Motivational reference group influence: Relationship to product complexity, conspicuousness and brand distinction. *European Research, 10*(2), 91–101.

Lewis, T. (2016). *Green Asia: Ecocultures, sustainable lifestyles, and ethical consumption*. New York, NY: Taylor & Francis.

Li, G., Li, G., & Kambele, Z. (2012). Luxury fashion brand consumers in China: Perceived value, fashion lifestyle, and willingness to pay. *Journal of Business Research, 65*(10), 1516–1522.

Liao, L. X., Corsi, A. M., Chrysochou, P., & Lockshin, L. (2015). Emotional response towards food packaging: A joint application of self-report and physiological measures of emotion. *Food Quality and Preference, 42*, 48–55. https://doi.org/10.1016/j.foodqual.2015.01.009

Limon, Y., Kahle, L. R., & Orth, U. R. (2009). Package design as a communications vehicle in cross-cultural values shopping. *Journal of International Marketing, 17*(1), 30–57.

Lin, Y.-C., & Wang, K.-Y. (2016). Language choice in advertising for multinational corporations and local firms: A reinquiry focusing on monolinguals. *Journal of Advertising, 45*(1), 43–52.

Lisa, W. (2007). Functional and symbolic attributes of product selection. *British Food Journal, 109*(2), 108–118. doi:10.1108/00070700710725482

Littel, S., & Orth, U. R. (2013). Effects of package visuals and haptics on brand evaluations. *European Journal of Marketing*. https://doi.org/10.1108/03090561311285510

Liu, W., Batra, R., & Wang, H. (2017). Product touch and consumers' online and offline buying: The role of mental representation. *Journal of Retailing, 93*(3), 369–381.

Locke, K. (2007). *Rational control and irrational free-play: Dual thinking modes as necessary tension in grounded theorizing*. Los Angeles, CA: Sage Publications.

Louviere, J., Islam, T., Wasi, N., Street, D., & Burgess, L. (2008). Designing discrete choice experiments: Do optimal designs come at a price? *Journal of Consumer Research, 35*(2), 360–375.

Low, G. S., & Lamb Jr, C. W. (2000). The measurement and dimensionality of brand associations. *Journal of Product & Brand Management, 9*(6), 350–370.

Luna, D., & Peracchio, L. A. (2005). Advertising to bilingual consumers: The impact of code-switching on persuasion. *Journal of Consumer Research, 31*(4), 760–765.

Lynch, J. G. (1991). *Memory and decision making handbook of consumer behavior* (1–9). Englewood Cliffs, NJ: Prentice-Hall.

Lynn, M. (1992). The psychology of unavailability: Explaining scarcity and cost effects on value. *Basic and Applied Social Psychology, 13*(1), 3–7.

Lynn, M., & Harris, J. (1997). The desire for unique consumer products: A new individual differences scale. *Psychology & Marketing, 14*(6), 601–616.

Lysonski, S. (2014). Receptivity of young Chinese to American and global brands: Psychological underpinnings. *Journal of Consumer Marketing, 31*(4), 250–262.

Ma, Q., Wang, M., & Da, Q. (2021). The effects of brand familiarity and product category in brand extension: An ERP study. *Neuroscience Research, 169*, 48–56.

Machiels, C. J., & Karnal, N. (2016). See how tasty it is? Effects of symbolic cues on product evaluation and taste. *Food Quality and Preference, 52*, 195–202. https://doi.org/10.1016/j.foodqual.2016.04.014

Machiels, C. J., & Orth, U. R. (2019). Multisensory packaging design across cultures. In C. Velasco & C. Spence (Eds.), *Multisensory packaging* (287–315). Cham: Palgrave Macmillan.

Madden, T., Hewett, K., & Roth, M. (2000). Managing images in different cultures: A cross-national study of color meanings and preferences. *Journal of International Marketing, 8*(4), 90–107.

Madzharov, A. V., & Block, L. G. (2010). Effects of product unit image on consumption of snack foods. *Journal of Consumer Psychology, 20*(4), 398–409.

Magnier, L., & Schoormans, J. (2015). Consumer reactions to sustainable packaging: The interplay of visual appearance, verbal claim and environmental concern. *Journal of Environmental Psychology, 44*, 53–62. https://doi.org/10.1016/j.jenvp.2015.09.005

Magnier, L., & Schoormans, J. (2017). How do packaging material, colour and environmental claim influence package, brand and product evaluations? *Packaging Technology and Science, 30*(11), 735–751.

Maio, G. R., & Esses, V. M. (2001). The need for affect: Individual differences in the motivation to approach or avoid emotions. *Journal of Personality, 69*(4), 583–614.

Malhotra, N. K. (2019). *Marketing research: An applied orientation* (7th ed.). Melbourne: Pearson.

Marian, L., Chrysochou, P., Krystallis, A., & Thøgersen, J. (2014). The role of price as a product attribute in the organic food context: An exploration based on actual purchase data. *Food Quality and Preference, 37*, 52–60. http://dx.doi.org/10.1016/j.foodqual.2014.05.001

Marlow, N., & Jansson-Boyd, C. V. (2011). To touch or not to touch; that is the question. Should consumers always be encouraged to touch products, and does it always alter product perception? *Psychology & Marketing, 28*(3), 256–266.

Marsh, K., & Bugusu, B. (2007). Food packaging—Roles, materials, and environmental issues. *Journal of Food Science, 72*(3), 39–55.

Marshall, C., & Rossman, G. B. (2014). *Designing qualitative research* (5th ed.). Thousand Oaks, CA: Sage Publications.

Martin, B. A. S., & Lee, C. K. C. (2004). The influence of ad model ethnicity and self-referencing on attitudes: Evidence from New Zealand. *Journal of Advertising, 33*(4), 27–37.

Mason, R. S. (1980). *Conspicuous consumption: A study of exceptional consumer behaviour.* Salford, CA: University of Salford.

Massa, S., & Testa, S. (2009). A knowledge management approach to organizational competitive advantage: Evidence from the food sector. *European Management Journal, 27*(2), 129–141.

McClure, S., Li, J., Tomlin, D., Cypert, K., Montague, L., & Montague, R. (2004). Neural correlates of behavioral preference for culturally familiar drinks. *Neuron, 44*(2), 379–387.

McKinsey & Company. (2019, 1 December). *The future of packaging.* www.mckinsey.com/industries/paper-forest-products-and-packaging/our-insights/the-future-of-packaging

McVee, M. B., Dunsmore, K., & Gavelek, J. R. (2005). Schema theory revisited. *Review of Educational Research, 75*(4), 531–566. doi:10.3102/00346543075004531

Media (2014, October). China overtook the US as the world's largest economy. *Business Insider Newspaper.* www.businessinsider.com.au/china-overtakes-us-as-worlds-largest-economy-2014-10?r=US&IR=T

Melendez, E. J., & Pritchard, R. H. (1985). Applying schema theory to foreign language reading. *Foreign Language Annals, 18*(5), 399–403. doi:10.1111/j.1944-9720.1985.tb00972.x

Melewar, T., & Vemmervik, C. (2004). International advertising strategy: A review, reassessment and recommendation. *Management Decision, 42*(7), 863–881.

Meyvis, T., & Janiszewski, C. (2004). When are broader brands stronger brands? An accessibility perspective on the success of brand extensions. *Journal of Consumer Research, 31*(2), 346–357.

Micu, C. C., & Coulter, R. A. (2010). Advertising in English in nonnative English-speaking markets: The effect of language and self-referencing in advertising in Romania on ad attitudes. *Journal of East-West Business, 16*(1), 67–84.

Middleton, C. H. (2015). International advertising: Using Hofstede's (1980) framework as a basis for standardization. *1996 Multicultural marketing conference, Virginia Beach, VA, USA.*

Mitchell, D., Moodie, C., Critchlow, N., & Bauld, L. (2019). Adolescents' perceptions of standardised cigarette packaging design and brand variant name post-implementation: A focus group study in Scotland. *BMC Public Health, 19*(1), 1227. https://doi.org/10.1186/s12889-019-7552-0

Monteiro, J., Silva, F. J. G., Ramos, S. F., Campilho, R. D. S. G., & Fonseca, A. M. (2019). Eco-design and sustainability in packaging: A survey. *Procedia Manufacturing, 38*, 1741–1749. https://doi.org/10.1016/j.promfg.2020.01.097

Morgan, C., Fajardo, T. M., & Townsend, C. (2021). Show it or say it: How brand familiarity influences the effectiveness of image-based versus text-based logos. *Journal of the Academy of Marketing Science, 49*(3), 566–583.

Moya, I., García-Madariaga, J., & Blasco, M.-F. (2020). What can neuromarketing tell us about food packaging? *Foods, 9*(12), 1856.

Mueller, S., & Lockshin, L. (2008). How important is wine packaging for consumers? On the reliability of measuring attribute importance with direct vebal versus indirect visual methods. *Paper presented at the 4th International Conference of the Academy of Wine Business Research, University of Siena, Siena.*

Mueller, S., Lockshin, L., & Louviere, J. (2010). What you see may not be what you get: Asking consumers what matters may not reflect what they choose. *Marketing Letters, 21*(4), 335–350.

Mueller, S., & Szolnoki, G. (2010). The relative influence of packaging, labelling, branding and sensory attributes on liking and purchase intent: Consumers differ in their responsiveness. *Food Quality and Preference, 21*(7), 774–783.

Murray, J. Y., Gao, G. Y., Kotabe, M., & Zhou, N. (2007). Assessing measurement invariance of export market orientation: A study of Chinese and non-Chinese firms in China. *Journal of International Marketing, 15*(4), 41–62.

Myers-Scotton, C. (1998). A theoretical introduction to the markedness model. In C. Myers-Scotton (Ed.), *Codes and consequences: Choosing linguistic varieties* (18–38). New York, NY: Oxford University Press.

Myers-Scotton, C. (2000). Explaining the role of norms and rationality in codeswitching. *Journal of Pragmatics, 32*, 1259–1271. https://doi.org/10.1016/S0378-2166(99)00099-5

Myers-Scotton, C. (2006). Natural codeswitching knocks on the laboratory door. *Bilingualism: Language and Cognition, 9*, 203–212.

Neave, L., Tzemou, E., & Fastoso, F. (2020). Seeking attention versus seeking approval: How conspicuous consumption differs between grandiose and vulnerable narcissists. *Psychology & Marketing, 37*(3), 418–427.

Neelankavil, J. P., Mummalaneni, V., & Sessions, D. N. (1995). Use of foreign language and models in print advertisements in East Asian countries: A logit modelling approach. *European Journal of Marketing, 29*(4), 24–38.

Neff, J. (2003). P&G outpacing unilever in five-year battle. *Advertising Age, 3*, 1–3.

Nelissen, R. M., & Meijers, M. H. (2011). Social benefits of luxury brands as costly signals of wealth and status. *Evolution and Human Behavior, 32*(5), 343–355.

Nemat, B., Razzaghi, M., Bolton, K., & Rousta, K. (2022). Design affordance of plastic food packaging for consumer sorting behavior. *Resources, Conservation and Recycling, 177*, 105949. https://doi.org/10.1016/j.resconrec.2021.105949

Newman, C. L., Burton, S., Andrews, J. C., Netemeyer, R. G., & Kees, J. (2018). Marketers' use of alternative front-of-package nutrition symbols: An examination of effects on product evaluations. *Journal of the Academy of Marketing Science, 46*(3), 453–476.

Ng, M., Chaya, C., & Hort, J. (2013). The influence of sensory and packaging cues on both liking and emotional, abstract and functional conceptualisations. *Food Quality and Preference, 29*(2), 146–156.

Nguyen, A. T., Parker, L., Brennan, L., & Lockrey, S. (2020). A consumer definition of eco-friendly packaging. *Journal of Cleaner Production, 252*, 119792. https://doi.org/10.1016/j.jclepro.2019.119792

Nguyen, H., & Gunasti, K. (2018). Original brands in competition against high quality copycats. *European Journal of Marketing, 52*(7/8), 1574–1597.

Nickels, W. G., & Jolson, M. A. (1976). Packaging, the fifth P in the marketing mix. *Advanced Management Journal, 41*(1), 13–26.

Nickerson, C., & Camiciottoli, B. C. (2013). Business English as a Lingua Franca in advertising texts in the Arabian gulf analyzing the attitudes of the Emirati community. *Journal of Business and Technical Communication, 27*(3), 329–352.
Oenning, E. J., Guimarães Junior, D., & Medeiros, D. D. D. (2018). Brazilian consumer preferences for processed tomato packaging using conjoint analysis. *Journal of Food Products Marketing, 24*(6), 724–744.
Olsen, M. C., Slotegraaf, R. J., & Chandukala, S. R. (2014). Green claims and message frames: How green new products change brand attitude. *Journal of Marketing, 78*(5), 119–137.
Ooijen, I. V. (2016). *The power of symbolic packaging cues advances in advertising research* (Vol. VI, 365–378). Wiesbaden: Springer Fachmedien Wiesbaden.
Orquin, J. L., Bagger, M. P., Lahm, E. S., Grunert, K. G., & Scholderer, J. (2019). The visual ecology of product packaging and its effects on consumer attention. *Journal of Business Research, 111*, 187–195.
Orth, U., & Malkewitz, K. (2006). Packaging design as resource for the construction of brand identity. *3rd International wine business research conference, Montpellier, France.*
Orth, U., & Malkewitz, K. (2008). Holistic package design and consumer brand impressions. *Journal of Marketing, 72*(3), 64–81. https://doi.org/10.1509/JMKG.72.3.064
Orth, U. R., & Crouch, R. C. (2014). Is beauty in the aisles of the retailer? Package processing in visually complex contexts. *Journal of Retailing, 90*(4), 524–537.
Otterbring, T., Rolschau, K., Furrebøe, E. F., & Nyhus, E. K. (2022). Crossmodal correspondences between typefaces and food preferences drive congruent choices but not among young consumers. *Food Quality and Preference, 96*, 104376. https://doi.org/10.1016/j.foodqual.2021.104376
Otterbring, T., Shams, P., Wästlund, E., & Gustafsson, A. (2013). Left isn't always right: Placement of pictorial and textual package elements. *British Food Journal, 115*(8), 1211–1225.
Özsomer, A. (2012). The interplay between global and local brands: A closer look at perceived brand globalness and local iconness. *Journal of International Marketing, 20*(2), 72–95.
Ozturk, A., & Cavusgil, S. T. (2019). Global convergence of consumer spending: Conceptualization and propositions. *International Business Review, 28*(2), 294–304.
Pakistan Country Guide (2018). *Pakistan – Labeling/marking requirements.* International Trade Administration. www.export.gov/article?id=Pakistan-Labeling-Marking-Requirements
Pakistan Economic Survey (2015), *Pakistan economic survey 2015-2016.* Government of Pakistan. www.finance.gov.pk/survey_1516.html
Paliwoda, S. J., Slater, S., Vrontis, D., Thrassou, A., & Lamprianou, I. (2009). International marketing adaptation versus standardisation of multinational companies. *International Marketing Review, 26*(4/5), 477–500.
Parasuraman, A., Grewal, D., & Krishnan, R. (2006). *Marketing research* (2nd ed.). Boston, MA: George T. Hoffman.
Park, W., & Lessig, P. (1981). Familiarity and its impact on consumer decision biases and heuristics. *Journal of Consumer Research, 8*(2), 223–231.
Peck, J., & Childers, T. L. (2006). If I touch it I have to have it: Individual and environmental influences on impulse purchasing. *Journal of Business Research, 59*(6), 765–769.
Peracchio, L. A., & Tybout, A. M. (1996). The moderating role of prior knowledge in schema-based product evaluation. *Journal of Consumer Research, 23*(3), 177–192.

Petty, R. E., & Cacioppo, J. T. (1986). In L. Berkowitz (Ed.), *The elaboration likelihood model of persuasion advances in experimental social psychology* (Vol. 19, 123–205). New York, NY: Academic Press.

Phau, I., & Suntornnond, V. (2006). Dimensions of consumer knowledge and its impacts on country of origin effects among Australian consumers: A case of fast-consuming product. *Journal of Consumer Marketing, 23*(1), 34–42.

Pieters, R., & Warlop, L. (1999). Visual attention during brand choice: The impact of time pressure and task motivation. *International Journal of Research in Marketing, 16*(1), 1–16.

Pietersen, W. (2010). Defining competitive advantage. *The European Business Review, Sep/Oct*, 34–39.

Pilditch, J. (1957). *The silent salesman.* London: Harper and Row.

Piñero, M. A., Lockshin, L., Kennedy, R., & Corsi, A. (2010). Distinctive elements in packaging (FMCG): An exploratory study. *Australian and New Zealand marketing academy conference.*

Piqueras-Fiszman, B., Velasco, C., & Spence, C. (2012). Exploring implicit and explicit crossmodal colour—Flavour correspondences in product packaging. *Food Quality and Preference, 25*(2), 148–155. http://dx.doi.org/10.1016/j.foodqual.2012.02.010

Piron, F. (2000). Consumers' perceptions of the country-of-origin effect on purchasing intentions of (in) conspicuous products. *Journal of Consumer Marketing, 17*(4), 308–321.

Podsakoff, P. M., MacKenzie, S. B., Lee, J.-Y., & Podsakoff, N. P. (2003). Common method biases in behavioral research: A critical review of the literature and recommended remedies. *Journal of Applied Psychology, 88*(5), 879.

Poon, P., Evangelista, F., & Albaum, G. (2010). Attitudes of migrants towards foreign-made products: An exploratory study of migrants in Australia. *Journal of Consumer Marketing, 27*(1), 35–42.

Ragaert, P., Verbeke, W., Devlieghere, F., & Debevere, J. (2004). Consumer perception and choice of minimally processed vegetables and packaged fruits. *Food Quality and Preference, 15*(3), 259–270.

Raggio, R. D., Leone, R. P., & Black, W. C. (2014). Beyond "halo": The identification and implications of differential brand effects across global markets. *Journal of Consumer Marketing, 31*(2), 133–144.

Raghubir, P., & Greenleaf, E. A. (2006). Ratios in proportion: What should the shape of the package be? *Journal of Marketing, 70*(2), 95–107. https://doi.org/10.1509/jmkg.70.2.095

Rahman, O., Zhu, X., & Liu, W. S. (2008). A study of the pyjamas purchasing behaviour of Chinese consumers in Hangzhou, China. *Journal of Fashion Marketing and Management: An International Journal, 12*(2), 217–231.

Rahman, R., Langner, T., & Temme, D. (2021). Brand love: Conceptual and empirical investigation of a holistic causal model. *Journal of Brand Management, 28*(6), 609–642.

Rao, A. R., & Monroe, K. B. (1988). The moderating effect of prior knowledge on cue utilization in product evaluations. *Journal of Consumer Research, 15*(2), 253–264.

Rathore, H., Jakhar, S. K., Kumar, S., & Kumar, M. E. (2022). Pay-what-you-want versus pick-your price: The interplay between participative pricing strategies and consumer's need for cognition. *Journal of Business Research, 141*, 73–84.

Razdan, R., Yamakawa, N., Francois, M., & Seong, J. (2021, 30 November). *The shifting consumer packaged goods market in a diversifying Asia*. McKinsey and Co. www.mckinsey.com/featured-insights/future-of-asia/the-shifting-consumer-packaged-goods-market-in-a-diversifying-asia

Reimann, M., Zaichkowsky, J., Neuhaus, C., Bender, T., & Weber, B. (2010). Aesthetic package design: A behavioral, neural, and psychological investigation. *Journal of Consumer Psychology, 20*(4), 431–441. https://doi.org/10.1016/j.jcps.2010.06.009

Reister, E., & Leidy, H. (2020). The effects of snack exposure, package size, and variety on subsequent energy intake in healthy adults. *Current Developments in Nutrition, 4*(Supplement_2), 1675–1675.

Rettie, R., & Brewer, C. (2000). The verbal and visual components of package design. *Journal of Product & Brand Management, 9*(1), 56–70.

Rigby, D. K., & Vishwanath, V. (2006). Localization—The revolution in consumer markets. *Harvard Business Review, 84*(4), 82–92.

Rind, H. M. (2015). *Pakistan proving irresistible to Western brands*. https://asia.nikkei.com/Economy/Pakistan-proving-irresistible-to-Western-brands

Rindfleisch, A., & Inman, J. (1998). Explaining the familiarity-liking relationship: Mere exposure, information availability, or social desirability? *Marketing Letters, 9*(1), 5–19.

Robertson, G. L. (2005). *Food packaging: Principles and practice* (2nd ed.). Boca Raton, FL: CRC Press.

Robinson, T., Borzekowski, D., Matheson, D., & Kraemer, H. (2007). Effects of fast food branding on young children's taste preferences. *Archives of Pediatrics & Adolescent Medicine, 161*(8), 792–797.

Roehm, M. L., & Brady, M. K. (2007). Consumer responses to performance failures by high-equity brands. *Journal of Consumer Research, 34*(4), 537–545.

Romaniuk, J. (2018). *Building distinctive brand assets*. Melbourne: Oxford University Press.

Romaniuk, J., & Nenycz-Thiel, M. (2013). Behavioral brand loyalty and consumer brand associations. *Journal of Business Research, 66*(1), 67–72.

Romaniuk, J., & Sharp, B. (2015). *How brands grow part 2: Including emerging markets, services, durables, new and luxury brands*. Adelaide: Oxford University Press.

Rompay, T. J., & Veltkamp, M. (2014). Product packaging metaphors: Effects of ambiguity and explanatory information on consumer appreciation and brand perception. *Psychology & Marketing, 31*(6), 404–415.

Roy Chaudhuri, H., Mazumdar, S., & Ghoshal, A. (2011). Conspicuous consumption orientation: Conceptualisation, scale development and validation. *Journal of Consumer Behaviour, 10*(4), 216–224.

Rundh, B. (2005). The multi-faceted dimension of packaging: Marketing logistic or marketing tool? *British Food Journal, 107*(9), 670–684.

Salciuviene, L., Ghauri, P. N., Salomea Streder, R., & De Mattos, C. (2010). Do brand names in a foreign language lead to different brand perceptions? *Journal of Marketing Management, 26*(11–12), 1037–1056. doi:10.1080/0267257X.2010.508976

Samiee, S. (2019). Reflections on global brands, global consumer culture and globalization. *International Marketing Review, 36*(4), 536–544.

Samli, A. C., Wilkinson, W. C., & Mutscheller, P. M. (2015). Successful globalization of parochial products. *Proceedings of the 1990 Academy of Marketing Science (AMS) annual conference*.

Sample, K. L., Hagtvedt, H., & Brasel, S. A. (2020). Components of visual perception in marketing contexts: A conceptual framework and review. *Journal of the Academy of Marketing Science, 48*(3), 405–421.

Santello, M. (2013). Advertising to Italian English bilinguals in Australia: Attitudes and response to language selection. *Applied Linguistics, 36*(1), 95–120.

Sapir, E. (1949). The unconscious patterning of behavior in society. In D. E. Mandelbaum (Ed.), *Selected writings of edward sapir in language, culture and personality* (554–559). Berkeley, CA: University of California Press.

Sattlegger, L. (2021). Making food manageable–packaging as a code of practice for work practices at the supermarket. *Journal of Contemporary Ethnography, 50*(3), 341–367. https://doi.org/10.1177/0891241620977635

Schifferstein, H. N. J., Lemke, M., & de Boer, A. (2022). An exploratory study using graphic design to communicate consumer benefits on food packaging. *Food Quality and Preference, 97*, 104458. https://doi.org/10.1016/j.foodqual.2021.104458

Schmitt, B. (1999). *Experiential marketing: How to get customers to sense, feel, think, act and relate to your company and brand.* New York, NY: The Free Press.

Schuldt, J. P. (2013). Does green mean healthy? Nutrition label color affects perceptions of healthfulness. *Health Communication, 28*(8), 814–821. https://doi.org/10.1080/10410236.2012.725270

Scott, M. L., Nowlis, S. M., Mandel, N., & Morales, A. C. (2008). The effects of reduced food size and package size on the consumption behavior of restrained and unrestrained eaters. *Journal of Consumer Research, 35*(3), 391–405.

Sevilla, J., & Kahn, B. E. (2014). The completeness heuristic: Product shape completeness influences size perceptions, preference, and consumption. *Journal of Marketing Research, 51*(1), 57–68. https://doi.org/10.1509/jmr.12.0153

Sharp, B. (2010). *How brands grow: What marketers don't know.* Melbourne: Oxford University Press.

Sheth, J. N. (2011). Impact of emerging markets on marketing: Rethinking existing perspectives and practices. *Journal of Marketing, 75*(4), 166–182.

Shiv, B., & Fedorikhin, A. (1999). Heart and mind in conflict: The interplay of affect and cognition in consumer decision making. *Journal of Consumer Research, 26*(3), 278–292.

Shove, E., & Warde, A. (2002). Inconspicuous consumption: The sociology of consumption, lifestyles and the environment. *Sociological Theory and the Environment: Classical Foundations, Contemporary Insights, 230*, 51.

Sielicka-Różyńska, M., Jerzyk, E., & Gluza, N. (2021). Consumer perception of packaging: An eye-tracking study of gluten-free cookies. *International Journal of Consumer Studies, 45*(1), 14–27.

Silayoi, P., & Speece, M. (2004). Packaging and purchase decisions: An exploratory study on the impact of involvement level and time pressure. *British Food Journal, 106*(8), 607–628.

Silayoi, P., & Speece, M. (2007). The importance of packaging attributes: A conjoint analysis approach. *European Journal of Marketing, 41*(11), 1495–1517.

Simmonds, G., & Spence, C. (2017). Thinking inside the box: How seeing products on, or through, the packaging influences consumer perceptions and purchase behaviour. *Food Quality and Preference, 62*, 340–351. https://doi.org/10.1016/j.foodqual.2016.11.010

Sivanathan, N., & Pettit, N. C. (2010). Protecting the self through consumption: Status goods as affirmational commodities. *Journal of Experimental Social Psychology, 46*(3), 564–570.

Smith, V., Barratt, D., & Sørensen, H. S. (2015). Do natural pictures mean natural tastes? Assessing visual semantics experimentally. *Cognitive Semiotics, 8*(1), 53–86. https://doi.org/10.1515/cogsem-2015-0001

Snyder, C. R. (1992). Product scarcity by need for uniqueness interaction: A consumer catch-22 carousel? *Basic and Applied Social Psychology, 13*(1), 9–24.

Sojka, J. Z., & Giese, J. L. (1997). Thinking and/or feeling: An examination of interaction between processing styles. *Advances in Consumer Research, 24*, 438–442.

Solomon, M. R. (1983). The role of products as social stimuli: A symbolic interactionism perspective. *Journal of Consumer Research, 10*(3), 319–329.

Souiden, N., Abdelaziz, F. B., & Fauconnier, A. (2013). Nutrition labelling: Employing consumer segmentation to enhance usefulness. *Journal of Brand Management, 20*(4), 267–282.

Spence, C. (2016). Multisensory packaging design: Color, shape, texture, sound, and smell. In P. Burgess (Ed.), *Integrating the packaging and product experience in food and beverages: A road-map to consumer satisfaction* (1–22). Duxford: Woodhead Publishing.

Spence, C., & Velasco, C. (2018). On the multiple effects of packaging colour on consumer behaviour and product experience in the 'food and beverage' and 'home and personal care' categories. *Food Quality and Preference, 68*, 226–237.

Spence, C., Velasco, C., & Petit, O. (2019). The consumer neuroscience of packaging. In C. Velasco & C. Spence (Eds.), *Multisensory packaging: Designing new product experiences* (319–347). Cham: Springer International Publishing.

Statistica (2019, 6 July). *Distribution of retail sales value of liquid detergents in China in 2019, by brand.* www.statista.com/statistics/1106878/china-liquid-detergent-retail-sales-value-share-by-brand/

Statistica (2021, 12 September). *Median forecast for China's GDP growth rate due to coronavirus COVID-19 pandemic in 2021 and 2022 as of October 2021.* Statistica. www.statista.com/statistics/1102691/china-estimated-coronavirus-covid-19-impact-on-gdp-growth/

Steckel, J. H., Winer, R. S., Bucklin, R. E., Dellaert, B. G., Drèze, X., Häubl, G., . . . Montgomery, A. L. (2005). Choice in interactive environments. *Marketing Letters, 16*(3–4), 309–320.

Steenkamp, J.-B. E., Batra, R., & Alden, D. L. (2003). How perceived brand globalness creates brand value. *Journal of International Business Studies, 34*(1), 53–65.

Steenkamp, J.-B. E., Van Heerde, H. J., & Geyskens, I. (2010). What makes consumers willing to pay a price premium for national brands over private labels? *Journal of Marketing Research, 47*(6), 1011–1024.

Stilley, K. M., Inman, J. J., & Wakefield, K. L. (2010). Spending on the fly: Mental budgets, promotions, and spending behavior. *Journal of Marketing, 74*(3), 34–47.

Street, D. J., Burgess, L., Viney, R., & Louviere, J. (2008). *Designing discrete choice experiments for health care.* Dordrecht: Springer.

Strizhakova, Y., & Coulter, R. A. (2015). Drivers of local relative to global brand purchases: A contingency approach. *Journal of International Marketing, 23*(1), 1–22.

Strizhakova, Y., & Coulter, R. A. (2019). Consumer cultural identity: Local and global cultural identities and measurement implications. *International Marketing Review, 36*(5), 610–627.

Sugermeyer, K. (2021, 2 December). *From conscious to non-conscious: Understanding the role of packaging design*. NielsenIQ. https://nielseniq.com/global/en/insights/education/2021/from-conscious-to-non-conscious-understanding-the-role-of-packaging-design/

Sullivan, O., & Gershuny, J. (2004). Inconspicuous consumption: Work-rich, time-poor in the liberal market economy. *Journal of Consumer Culture, 4*(1), 79–100.

Tai, S. H. (1997). Advertising in Asia: Localize or regionalize? *International Journal of Advertising, 16*(1), 48–61.

Tai, S. H., & Pae, J. H. (2002). Effects of TV advertising on Chinese consumers: Local versus foreign-sourced commercials. *Journal of Marketing Management, 18*(1–2), 49–72.

Tajfel, H. (1978). Social categorization, social identity and social comparison. In H. Tajfel (Ed.), *Differentiation between social groups* (66–76). London: Academic Press.

Tan, Q., & Sousa, C. M. (2013). International marketing standardization. *Management International Review, 53*(5), 711–739.

Taylor, S. E., & Fiske, S. T. (1975). Point of view and perceptions of causality. *Journal of Personality and Social Psychology, 32*(3), 439–445.

Thomas, A. (2000). Elements influencing wine purchasing: A New Zealand view. *International Journal of Wine Marketing, 12*(2), 47–62.

Thompson, D. V., Hamilton, R. W., & Rust, R. T. (2005). Feature fatigue: When product capabilities become too much of a good thing. *Journal of Marketing Research, 42*(4), 431–442.

Timmermans, S., & Tavory, I. (2012). Theory construction in qualitative research: From grounded theory to abductive analysis. *Sociological Theory, 30*(3), 167–186.

Toffoli, R., & Laroche, M. (2002). Cultural and language effects on Chinese bilinguals' and Canadians' responses to advertising. *International Journal of Advertising, 21*(4), 505–524.

Tonby, O., Woetzel, J., Choi, W., Seong, J., & Wang, P. (2019, 1 October). *Asia's future is now*. McKinsey Global Institute. www.mckinsey.com/featured-insights/asia-pacific/five-windows-of-opportunity-for-postpandemic-asia

Tonby, O., Woetzel, J., Razdan, R., Choi, W., Smit, S., Yamakawa, N., . . . Devesa, T. (2021, 1 December). *Beyond income: Redrawing Asia's consumer map*. McKinsey Global Institute. www.mckinsey.com/featured-insights/asia-pacific/beyond-income-redrawing-asias-consumer-map

Torres, I. M., & Luna-Nevarez, C. (2012). What products can benefit from African American advertising appeals? The moderating role of product involvement. *Journal of Current Issues & Research in Advertising, 33*(1), 37–55.

Trading-Economics. (2015). *Pakistan GDP growth rate*. www.tradingeconomics.com/pakistan/gdp-growth

Train, K. E. (2009). *Discrete choice methods with simulation*. Cambridge: Cambridge University Press.

Tu, L. L., Kwon, J., & Gao, H. (2021). Heart or mind? The impact of power distance belief on the persuasiveness of cognitive versus affective appeal in education marketing messages. *Journal of Marketing Research*. https://doi.org/10.1177/00222437211002196

Tu, Y., Yang, Z., & Ma, C. (2015). Touching tastes: The haptic perception transfer of liquid food packaging materials. *Food Quality and Preference, 39*(January), 124–130.

Turner, J. C. (1982). Towards a cognitive redefinition of the social group. In H. Tajfel (Ed.), *Social identity and intergroup relations* (15–40). Cambridge: Cambridge University Press.

Turner, J. C., Brown, R. J., & Tajfel, H. (1979). Social comparison and group interest in ingroup favouritism. *European Journal of Social Psychology, 9*(2), 187–204.

Tybout, A. M., Calder, B. J., & Sternthal, B. (1981). Using information processing theory to design marketing strategies. *Journal of Marketing Research, 18*(1), 73–79.

UN DESA. (2019, 5 October). *World population prospects 2019.* United Nations Department of Economic and Social Affairs. https://population.un.org/wpp/Publications/Files/WPP2019_Volume-I_Comprehensive-Tables.pdf

Underwood, R. L. (2003). The communicative power of product packaging: Creating brand identity via lived and mediated experience. *Journal of Marketing Theory and Practice, 11*(1), 62–76.

Underwood, R. L., & Klein, N. M. (2002). Packaging as brand communication: Effects of product pictures on consumer responses to the package and brand. *Journal of Marketing Theory and Practice, 10*(4), 58–68. https://doi.org/10.1080/10696679.2002.11501926

Underwood, R. L., Klein, N. M., & Burke, R. R. (2001). Packaging communication: Attentional effects of product imagery. *Journal of Product & Brand Management, 10*(7), 403–422.

Unni, R., & Pillai, D. (2010). Context specificity in use of price information sources. *Journal of Consumer Marketing, 27*(3), 243–250.

Urbany, J. E., Dickson, P. A., & Kalapurakal, R. (1996). Price search in the retail grocery market. *Journal of Marketing, 60*(2), 91–104.

Valentine, V., & Gordon, W. (2000). The 21st century consumer: A new model of thinking. *International Journal of Market Research, 42*(2), 185–206.

Van den Berg-Weitzel, L., & Van de Laar, G. (2001). Relation between culture and communication in packaging design. *The Journal of Brand Management, 8*(3), 171–184.

Van Ooijen, I. (2016). The power of symbolic packaging cues. In P. Verlegh, H. Voorveld, & M. Eisend, (Eds.), *Advances in advertising research* (Vol. VI, 365–378). Wiesbaden: European Advertising Academy. Springer Gabler..

Van Ooijen, I., Fransen, M. L., Verlegh, P. W., & Smit, E. G. (2017). Signalling product healthiness through symbolic package cues: Effects of package shape and goal congruence on consumer behaviour. *Appetite, 109*, 73–82.

Van Rompay, T. J., De Vries, P. W., Bontekoe, F., & Tanja-Dijkstra, K. (2012). Embodied product perception: Effects of verticality cues in advertising and packaging design on consumer impressions and price expectations. *Psychology & Marketing, 29*(12), 919–928.

Van Rompay, T. J. L., Fransen, M. L., & Borgelink, B. G. D. (2014). Light as a feather: Effects of packaging imagery on sensory product impressions and brand evaluation. *Marketing Letters, 25*(4), 397–407. https://doi.org/10.1007/s11002-013-9260-3

Veblen, T. (1899). *The theory of the leisure class.* New Brunswick, NJ: Transaction Publishers.

Velasco, C., Salgado-Montejo, A., Marmolejo-Ramos, F., & Spence, C. (2014). Predictive packaging design: Tasting shapes, typefaces, names, and sounds. *Food Quality and Preference, 34*, 88–95.

Velasco, C., & Spence, C. (2018). *Multisensory packaging: Designing new product experiences.* Cham: Springer.

Velasco, C., & Spence, C. (2019). The role of typeface in packaging design. In C. Velasco & C. Spence (Eds.), *Multisensory packaging: Designing new product experiences* (79–101). Cham: Springer International Publishing.

Velasco, C., Wan, X., Salgado-Montejo, A., Woods, A., Oñate, G. A., Mu, B., & Spence, C. (2014). The context of colour–flavour associations in crisps packaging: A cross-cultural study comparing Chinese, Colombian, and British consumers. *Food Quality and Preference*, *38*, 49–57.

Velasco, C., Woods, A. T., Hyndman, S., & Spence, C. (2015). The taste of typeface. *i-Perception*, *6*(4), 2041669515593040. https://doi.org/10.1177/2041669515593040

Verplanken, B. (1991). Persuasive communication of risk information: A test of cue versus message processing effects in a field experiment. *Personality and Social Psychology Bulletin*, *17*(2), 188–193.

Verplanken, B., Hazenberg, P. T., & Palenewen, G. R. (1992). Need for cognition and external information search effort. *Journal of Research in Personality*, *26*(2), 128–136.

Verplanken, B., & Pieters, R. G. (1988). Individual differences in reverse hindsight bias: I never thought something like Chernobyl would happen. Did I? *Journal of Behavioral Decision Making*, *1*(3), 131–147.

Vihma, S. (1995). *Products as representations: A semiotic and aesthetic study of design products*. Helsinki: University of Art and Design Helsinki.

Vila-López, N., Kuster-Boluda, I., & Alacreu-Crespo, A. (2021). Designing a low-fat food packaging: Comparing consumers' responses in virtual and physical shopping environments. *Foods*, *10*(2), 211. https://doi.org/10.3390/foods10020211

Voss, K. E., Spangenberg, E. R., & Grohmann, B. (2003). Measuring the hedonic and utilitarian dimensions of consumer attitude. *Journal of Marketing Research*, *40*(3), 310–320.

Vrontis, D. (2005). The creation of the AdaptStand process in international marketing. *Journal of Innovative Marketing*, *1*(2), 7–21.

Wagner, U., & Charinsarn, A. R. (2021). What language should be displayed on product packaging? How unconventional lettering influences packaging and product evaluation. *Journal of International Consumer Marketing*, *33*(1), 1–18. doi:10.1080/08961530.2020.1741483

Walton, D. (2004). *Abductive reasoning*. Tuscaloosa, AL: University of Alabama Press.

Wan, X., Velasco, C., Michel, C., Mu, B., Woods, A. T., & Spence, C. (2014). Does the type of receptacle influence the crossmodal association between colour and flavour? A cross-cultural comparison. *Flavour*, *3*(1), 3.

Ward, E., Yang, S., Romaniuk, J., & Beal, V. (2020). Building a unique brand identity: Measuring the relative ownership potential of brand identity element types. *Journal of Brand Management*, *27*(4), 393–407. https://doi.org/10.1057/s41262-020-00187-6

Warlop, L., Ratneshwar, S., & Van Osselaer, S. M. (2005). Distinctive brand cues and memory for product consumption experiences. *International Journal of Research in Marketing*, *22*(1), 27–44.

Weinberg, P., & Gottwald, W. (1982). Impulsive consumer buying as a result of emotions. *Journal of Business Research*, *10*(1), 43–57.

Wells, L. E., Farley, H., & Armstrong, G. A. (2007). The importance of packaging design for own-label food brands. *International Journal of Retail & Distribution Management*, *35*(9), 677–690. https://doi.org/10.1108/09590550710773237

Wertenbroch, K., & Skiera, B. (2002). Measuring consumers' willingness to pay at the point of purchase. *Journal of Marketing Research, 39*(2), 228–241.

White, K., Lin, L., Dahl, D. W., & Ritchie, R. J. (2016). When do consumers avoid imperfections? Superficial packaging damage as a contamination cue. *Journal of Marketing Research, 53*(1), 110–123.

Whorf, B. L. (1956). Language, mind, and reality. In J. B. Carroll (Ed.), *Language, thought, and reality: Selected writings of Benjamin Lee Whorf* (246–270). Cambridge, MA: MIT Press.

Williams, H., Lindström, A., Trischler, J., Wikström, F., & Rowe, Z. (2020). Avoiding food becoming waste in households—The role of packaging in consumers' practices across different food categories. *Journal of Cleaner Production, 265*, 121775. https://doi.org/10.1016/j.jclepro.2020.121775

Winit, W., Gregory, G., Cleveland, M., & Verlegh, P. (2014). Global vs local brands: How home country bias and price differences impact brand evaluations. *International Marketing Review, 31*(2), 102–128.

Woetzel, J., Tonby, O., Thompson, F., Burtt, P., & Lee, G. (2014, 10 September). *Southeast Asia at the crossroads: Three paths to prosperity.* McKinsey Global Institute. www.mckinsey.com/~/media/McKinsey/Featured%20Insights/Asia%20Pacific/Three%20paths%20to%20sustained%20economic%20growth%20in%20Southeast%20Asia/MGI%20SE%20Asia_Executive%20summary_November%202014.pdf

Wong, N., Rindfleisch, A., & Burroughs, J. E. (2003). Do reverse-worded items confound measures in cross-cultural consumer research? The case of the material values scale. *Journal of Consumer Research, 30*(1), 72–91. https://doi.org/10.1086/374697

Wood, L. (2007). Functional and symbolic attributes of product selection. *British Food Journal, 109*(2), 108–118.

Xu, X., & Chen, R. (2016). The role of a model's race in influencing Chinese consumers' product perception. *Asian Business & Management, 15*(3), 201–225.

Yan, D., Sengupta, J., & Wyer Jr, R. S. (2014). Package size and perceived quality: The intervening role of unit price perceptions. *Journal of Consumer Psychology, 24*(1), 4–17.

Yang, S., & Raghubir, P. (2005). Can bottles speak volumes? The effect of package shape on how much to buy. *Journal of Retailing, 81*(4), 269–281. https://doi.org/10.1016/j.jretai.2004.11.003

Yap, S.P.W., & Yazdanifard, R. (2014). Comparison on the impact of standardization and adaptation on international marketing. *Journal of Research in Marketing, 3*(1), 250–259.

Yarar, N., Machiels, C. J. A., & Orth, U. R. (2019). Shaping up: How package shape and consumer body conspire to affect food healthiness evaluation. *Food Quality and Preference, 75*, 209–219. https://doi.org/10.1016/j.foodqual.2019.03.004

Yendamuri, P., & Ingilizian, Z. (2019, 20 August). *In 2020 Asia will have the world's largest GDP. Here's what that means.* World Economic Forum. www.weforum.org/agenda/2019/12/asia-economic-growth/

Yetts, W. P. (1912). *Symbolism in Chinese art.* Leyden: E.J. Brill.

Yip, G. S., & Hult, G. T. M. (2012). *Total global strategy* (3rd ed.). Boston, MA: Pearson Prentice Hall.

Yoon, C., Gutchess, A. H., Feinberg, F., & Polk, T. A. (2006). A functional magnetic resonance imaging study of neural dissociations between brand and person judgments. *Journal of Consumer Research, 33*(1), 31–40.

YouGov. (2021, 16 November). *International FMCG/CPG Report 2021.* YouGov. www.yougov.com/fmcg

Young, S. (2002). Packaging design, consumer research, and business strategy: The march toward accountability. *Design Management Journal (Former Series), 13*(4), 10–14.

Zajonc, R. B. (1968). Attitudinal effects of mere exposure. *Journal of Personality and Social Psychology, 9*(2/Part 2), 1–27.

Zhan, L., & He, Y. (2012). Understanding luxury consumption in China: Consumer perceptions of best-known brands. *Journal of Business Research, 65*(10), 1452–1460.

Zhang, D. (2014). A cross-cultural exploration of children's preferences of package design: The US and China. *Journal of International Consumer Marketing, 26*(5), 391–404.

Zhou, L., & Hui, M. K. (2003). Symbolic value of foreign products in the People's Republic of China. *Journal of International Marketing, 11*(2), 36–58.

Zhou, L., Poon, P., & Wang, H. (2015). Consumers' reactions to global versus local advertising appeals: A test of culturally incongruent images in China. *Journal of Business Research, 68*(3), 561–568.

Zhou, L., & Wong, A. (2008). Exploring the influence of product conspicuousness and social compliance on purchasing motives of young Chinese consumers for foreign brands. *Journal of Consumer Behaviour: An International Research Review, 7*(6), 470–483.

Zhou, N., & Belk, R. W. (2004). Chinese consumer readings of global and local advertising appeals. *Journal of Advertising, 33*(3), 63–76.

Zhu, M., Billeter, D. M., & Inman, J. J. (2012). The double-edged sword of signaling effectiveness: When salient cues curb postpurchase consumption. *Journal of Marketing Research, 49*(1), 26–38.

Zucchella, A., & Magnani, G. (2016). *International entrepreneurship: Theoretical foundations and practices* (2nd ed.). Hampshire: Springer.

Index

Note: Page numbers in *italics* indicate figures and in **bold** indicate tables on the corresponding pages.